SUCCEEDING IN STATISTICS

A Review of Math, Spreadsheet, and Graphing Calculator Skills for Elementary Statistics

RONALD E. SHIFFLER
Western Carolina University

ARTHUR J. ADAMS
University of Louisville

Duxbury Press

An Imprint of Brooks/Cole Publishing Company

I(T)P® An International Thomson Publishing Company

Pacific Grove • Albany • Belmont • Bonn • Boston • Cincinnati • Johannesburg • London
Madrid • Melbourne • Mexico City • New York • Scottsdale • Singapore • Tokyo • Toronto

Sponsoring Editor: *Curt Hinrichs*
Marketing Team: *Jean Thompson, Laura Hubrich, and Deanne Brown*
Editorial Assistant: *Carrie Izant*
Production Editor: *Tom Novack*

Permissions Editor: *Carline Haga*
Cover Designer: *Roy R. Neuhaus*
Cover Printing: *Webcom Limited*
Printing and Binding: *Webcom Limited*

For more information, contact Duxbury Press at Brooks/Cole Publishing Company:

BROOKS/COLE PUBLISHING COMPANY
511 Forest Lodge Road
Pacific Grove, CA 93950
USA

International Thomson Publishing Europe
Berkshire House 168-173
High Holborn
London WC1V 7AA
England

Thomas Nelson Australia
102 Dodds Street
South Melbourne, 3205
Victoria, Australia

Nelson Canada
1120 Birchmount Road
Scarborough, Ontario
Canada M1K 5G4

International Thomson Editores
Seneca 53
Col. Polanco
11560 México, D. F., México

International Thomson Publishing GmbH
Königswinterer Strasse 418
53227 Bonn
Germany

International Thomson Publishing Asia
60 Albert Street
#15-01 Albert Complex
Singapore 189969

International Thomson Publishing Japan
Hirakawacho Kyowa Building, 3F
2-2-1 Hirakawacho
Chiyoda-ku, Tokyo 102
Japan

Printed in Canada

10   9   8   7   6   5   4   3   2   1

Library of Congress Cataloging-in-Publication Data
Shiffler, Ronald E.
  Succeeding in statistics: a review of math, spreadsheet, and graphing calculator skills for elementary statistics / Ronald E. Shiffler, Arthur J. Adams
        p.   cm.
    Includes index.
    ISBN 0-534-36234-6
    1. Mathematics.   I. Adams, Arthur J.   II. Title.
QA39.2.S496 1999
510—dc21   98-51474

TABLE OF CONTENTS

To the Student

Congratulations on your purchase of this book. This tells us that you are serious about brushing up on those areas of mathematics that are considered prerequisite to the statistics course in which you are currently enrolled.

We prepared this book with you in mind. We anticipate that you fit one of three student profiles that we have encountered in our statistics classes:

☞ You took the prerequisite math course(s) a *long time* ago and you feel certain that you have <u>forgotten</u> most of the material. We're talking <u>years</u> since you had a math course. You can probably remember all the words to a popular song that was playing when you took that math course; but, do you remember the math?

☞ You took the prerequisite math course(s) recently, but you did not have a good experience with the class(es). Either you didn't connect with your instructor(s) or you feel you just didn't learn anything for a variety of reasons.

☞ You are feeling a bit insecure about your math background. You may or may not have taken the prerequisite math course(s). However, math has never been your favorite subject, so you're not sure if you're <u>really</u> prepared for this statistics course or not.

Now that we have established that your math skills are not as sharp as you'd like, what can we do about it? Together we hope you will work through this book at your own pace to fulfill your special needs. You may not need to study every chapter in this book. If you have the time to work through the whole book, we strongly encourage you to do so. If not, focus on the one or two chapters in this book that most relate to your current statistics material. In a very rough sense the chapters in this book fall into one of two categories: <u>fundamental</u> arithmetic or algebra topics or <u>intermediate</u> topics that directly tie into almost all statistics courses. In terms of the chapters included within each category consider the following flow chart:

```
╔══════════════════════════╗
║      Fundamental         ║
║     Chapters 1, 4, 8     ║
╚══════════════════════════╝
```

```
╔══════════════════════════╗
║      Intermediate        ║
║   Chapters 2, 3, 5, 6, 7 ║
╚══════════════════════════╝
```

Everyone should read and study the fundamental chapters (1, 4, and 8). The remaining intermediate chapters (2, 3, 5, 6, and 7) could be read on a just-in-time basis.

At the beginning of the book is a chart that cross references the chapter material in this book with the material in your current statistics book. You will have to bear with us a bit as we can't know exactly which chapter in your statistics book corresponds to our material. (There are a lot of statistics books out there!) We've tried to show you the tie-in using major topical headings. It's up to you to translate the topical headings into a chapter number in your text.

To use the material in this book we recommend several approaches - Plans A, B, and C. Each chapter opens with a pretest and closes with a posttest; the answers to all questions are in the back.

Plan A involves starting the chapter by taking the pretest to see how much you really know. Point values for each question and a time limit are included with each pretest. Each test is self-graded; don't be too easy on yourself! Only give yourself a point if you get the correct answer. To be up to speed with the chapter material, you should either have a perfect score on the pretest or miss at most 1 point.

If you score this well, immediately go to the posttest at the end of the chapter and take it. Again, if you make a perfect score (or are within a point of a perfect score), then you should skip the chapter because you are about as proficient with that chapter material as you can get.

On the other hand, if you don't ace the pretest and/or the posttest, then there is room for improvement. We suggest Plan B. If you have a tendency to be a bit impatient, or you don't wish to read all of the text in each chapter, look for the Summary Examples in each chapter. Read and study each of these Summary Examples and then try to solve the chapter exercises. The answers are in the back. Work as many exercises as needed until you feel comfortable with the material (and you consistently get the correct answers to the exercises). To validate your competency with the material you might take the posttest again. Presumably you should earn a higher score this time around.

If you still don't feel really sure of yourself, then go to Plan C, which means you should read the entire chapter -- text and all boxed material -- and work all the chapter exercises before tackling the posttest.

In summary, here are the options for tackling each chapter:

HOW TO STUDY THE MATERIAL IN EACH CHAPTER

PLAN	IF YOU ARE THIS TYPE OF PERSON	DO THIS	HOW TO JUDGE YOUR MASTERY OF THE MATERIAL
A	Impatient	Take the chapter pretest and posttest.	Earn a perfect score (or close to it) on each test.
B	Prone to make careless errors	Take the chapter pretest. Then read all the Summary Examples and work some of the chapter exercises. Take the posttest.	You get the correct answers for most of the chapter exercises and you ace the posttest.
C	Slow but steady	Take the chapter pretest. Then read and work through the entire chapter and the chapter exercises. Take the posttest.	You ace the posttest.

EXAMPLES FOR TI-83® and EXCEL® USERS

Most of the examples shown in this book are presented in three ways:

① By hand computation   ② By TI-83 calculator   ③ By Excel spreadsheet

The primary emphasis is on doing problems by hand, but if you have access to a TI-83 or similar model and/or Excel, you may wish to see how to generate solutions using these tools. We will usually show the hand computation first, then follow with steps for the TI-83 and Excel results. See Chapter 0 for preliminary instructions using the TI-83 and Excel. For each Excel application we have summarized the keystrokes and point-and-click sequences needed to produce a desired outcome. These instructions are found under headings labeled Excel F.Y.I.

HAVE FUN AND GOOD LUCK!

We would like to recognize the extraordinary efforts of our typist, Cindy McDonald, who created all the beautiful pages in this book; our editor, Curt Hinrichs, who saw the wisdom in publishing a just-in-time book for you; our production editor, Tom Novack, and Carrie Izant, editorial assistant, who handled all the details to make the book publishing process smooth and trouble free. Also the following reviewers supplied <u>many</u> good tips and suggestions for improvement:

Ramakant Khazanie, Humboldt State University
Dennis Kimzey, Rogue Community College
Bruce Trumbo, California State University -- Hayward
Paul Paschke, Oregon State University

<u>Thanks, gang!</u>

CHAPTER ZERO
TI-83 and Excel Basics

(Skip this chapter if you are already familiar with the TI-83 and/or Excel, or if you simply want to learn the various concepts by hand.)

0.1   TI-83 Preliminaries

The TI-83 keyboard looks intimidating at first. Sure, you know what some of the keys are for, but you have no clue on lots of the other ones. And where is the "equal sign" anyway? Not to worry. We'll not need all the keys, and we'll explain the ones we do need as we go along through the following chapters.

When we wish to illustrate an operation that involves a series of keystrokes, we will adopt the following convention in this book: *we will list the keys to press, separated by commas.* For instance, turn on your TI-83 and then press 8, X, 9, ENTER. This sequence multiplies 8 times 9: you should have 72 showing on the screen. It should now be apparent that the ENTER key is also your "equal sign" key. In the following sequence, "(" is the left parenthesis, located above the 8 key; ")" is the right parenthesis, above the 9 key. Try this: 5, (, 3, +, 8,), -, 9, ENTER. The result should be 46.

Here are a few special purpose keys that we will encounter often when using the TI-83 calculator (refer to the picture of the TI-83 keypad on the next page):

❶   **The CLEAR key.** Use this key to clear the screen; to start over fresh, for instance, if you've messed up a line or an equation.

❷   **The negation (-) key, below the 3 key.** Use this key to enter a negative number. For instance, to multiply negative 4 times 6, press (-), 4, X ,6, ENTER; you should display -24. Do <u>not</u> use the blue subtraction key to enter a negative number. If you did so in the previous example, you would either get an error message or you would subtract 24 from the result of the most recent calculation.

❸   **The green ALPHA key.** When you press ALPHA, the letter printed in green above certain keys will appear on the screen on the next keystroke. For instance, press ALPHA, P (above the 8) to get the letter P on the screen. If we wished to label a variable as P, or X, or Y, etc., we would use the ALPHA key to do so.

❹   **The four direction keys.** These are the blue keys with arrows on them. Use them on menu screens to move up, down, left, or right. The left arrow key also serves as a backspace.

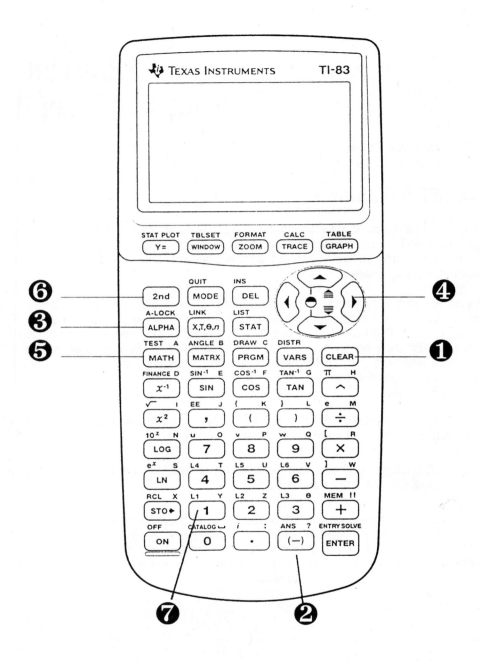

Material extracted from TI-83 Guidebook
with permission of the publisher
Copyright 1996, Texas Instruments Incorporated

❺ **The MATH key.** This key produces a menu -- choices of operations we may wish to pick from. To illustrate, press MATH and you'll see some choices. If we want option 4, cube root, press 4 to activate it. For instance, to evaluate $\sqrt[3]{512}$, the cube root of 512, press MATH, 4, 512, ENTER. Your display should show the answer: 8. As another example, if you wished to evaluate 8! (which is 8-factorial, or 8x7x6x5x4x3x2x1), press 8, MATH, left arrow, 4, ENTER. Do you have 40,320?

❻ **The yellow 2nd key.** Most keys have two functions. The second function is printed in yellow above each key. For instance the x^2 key has a second function: it is also used to obtain a square root. Need the square root of 643.1296? Press 2nd, $\sqrt{}$, 643.1296, ENTER. Do you have 25.36?

❼ **The L1 (List 1) key.** When we enter a data set into the TI-83, the number will be stored in a "list." The TI-83 can store several lists using list names like L2, L3, etc. We'll see later that you can give a data set any name you wish instead of the generic L1, L2, etc. designation. Note that use of the L1 key would be immediately preceded by activating the 2nd function key.

0.2 Excel Preliminaries

Introduction

Your ability to use a computer to master the material in this book will depend on your familiarity with the Excel program. As you gain confidence with Excel you will find that some of the difficulties that you might have experienced previously with mathematics literally disappear right before your eyes! Not only will you learn how to use a spreadsheet application -- Excel -- but also you will refresh your math skills needed for future success in your statistics course.

Throughout the book we will include summaries of Excel instructions under headings called "EXCEL FYI." To maximize your learning potential, we recommend that you read the text first and simultaneously follow along at your computer. Thereafter, you may find it easier to consult the EXCEL FYI boxes for quick recall.

Welcome to Excel

In this section we intend to show you how to "get into" the Excel program. Part of the mystique of using computers is the jargon that computer people use. They seem to use words and phrases that we may not understand. Putting some of these phrases together in sentences creates a "foreign" language for the non-computer literate person. For example, Excel is called a "spreadsheet" program because it resembles the old Wilson Jones worksheets that accountants used to create budgets by hand with pencil and paper.

To begin let's open the Excel application. We will assume that you have turned on your computer and your screen displays the standard desktop image. Because there are a variety of ways to open Excel, we hesitate to go into much detail here. If you have never used Excel before, try the following sequence: Click on the **Start** button in the lower left-hand corner of the screen. Highlight the word **Programs** to bring up a menu of choices. (What we mean by "highlight" is that you should move your mouse until the white arrow points to the word **Programs**, which should cause a rectangular section of your screen surrounding the word **Programs** to turn a darker color.)

If the phrase **Microsoft Excel** appears on the menu, click on this phrase and wait until the following figure appears on your screen. (If the phrase **Microsoft Excel** does not appear on your screen, look for the phrase **Office 97**. Highlighting this phrase should produce a submenu with **Microsoft Excel** on it. Click on **Microsoft Excel**.)

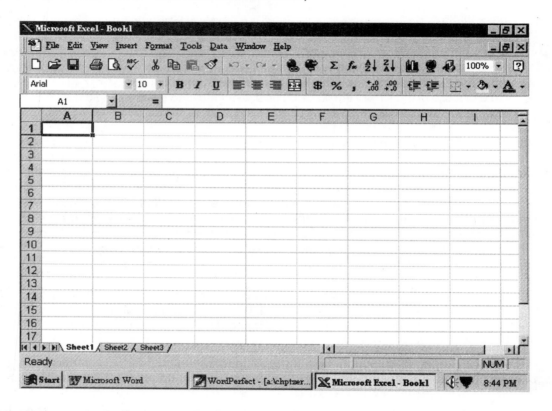

EXCEL F.Y.I.

To open Excel:

     (1) turn on your computer,
     (2) click on the **Start** button in the lower left-hand corner,
     (3) highlight the word **Programs**,
     (4) click on the phrase **Microsoft Excel** (or highlight the phrase **Office 97** and
          then click on **Microsoft Excel**).

Spreadsheet Basics

Obviously, one of the first things you notice about the preceding figure is the grid that covers up most of the screen. The sequences of horizontal and vertical lines create "cells" on the screen. Each cell has an "address," which refers to its column position and its row position. Columns are labeled with capital letters: A, B, C, and so on. Rows are labeled with numbers: 1, 2, 3, and so on. Thus, location "C3" refers to the cell formed by crossing column C with row 3.

Look closely at the top of the figure. In the first horizontal line you should see the Excel icon and then the phrase "Microsoft Excel - Book1." The top line tells you what program you are currently using and then the name of the file. At this point the name of our file is "Book1," but we will change that name later.

The second horizontal line from the top contains short names like File, Edit, View, Insert, and so on, reading from left to right. This line is called the **menu bar** and beneath each word is a hidden menu of options. When you point at one of these words with the white arrow and then click, the menu appears. (Pointing at a word or symbol on the screen with the white arrow and then clicking is called **selecting** that word or symbol.)

The third and fourth horizontal lines from the top of the screen are called "toolbars." The third line is called the **standard toolbar** and the fourth line is called the **formatting toolbar**. On each toolbar you will see small shapes that coincide with specific commands. For instance, on the standard toolbar you should see the image of a pair of scissors, which represents the command **cut**, while on the formatting toolbar you should see a capital **B** letter, which represents the command **bold**. Point at these symbols with the white arrow and let the arrow rest on each symbol for a moment without clicking. The words "Cut" and "Bold" will appear.

Your screen is capable of showing several other toolbars as you work, or you can decide to remove the toolbars from your screen. Our suggestion is to leave them on your screen until you gain more familiarity with Excel. In our figure, we have four toolbars **activated**, which means that they are displayed around the edges of our main Excel spreadsheet screen. The four toolbars are: standard, formatting, formula, and status. (To add or remove toolbars from your screen you can use the View button on the second horizontal line.)

If you have been paying close attention to the screen, you will have noticed that the "white arrow" that moves around the screen as you move your mouse takes on different shapes. When we move it around the top of the screen from one toolbar to another, the arrow shape appears. But when you move your mouse so that the shape is in the middle of the spreadsheet, the white arrow changes to a white cross. In the process of moving around the screen you might also have noticed three or four other shapes that suddenly appear, depending on where you are on the screen. For now, we need to concern ourselves with only the arrow shape and the cross shape.

Move the white cross to cell C4 and click. This is called making cell C4 **active**. What you should see on your screen is a black border encircling the perimeter of cell C4. Also, the "C" in the column heading is slightly raised as is the "4" in the row heading. Also notice the "C4" designation that appears as the first entry in the formula toolbar line (right above the letter "A" in the column headings).

The raised symbols in the row and column headings tell you what cell you are working in. The same information is shown in the formula toolbar line. As obvious as this may seem, you will sometimes find yourself wondering what cell you're in when working on more involved spreadsheet applications. When in doubt, look for the raised symbols in the row and column headings or look in the formula toolbar.

Entering Data in an Excel Spreadsheet

Our next session with Excel involves data entry. Suppose we want to enter the data set that appears below into an Excel spreadsheet:

Type of Bedding	Number Sold
Single	8
Double	12
Queen	9
King	5

Notice that the first column of information is textual, while the second column is numerical.

If cell A1 is not already active, make it active by positioning the white cross on that cell and clicking once. Now use the keyboard to key in the word "Type." (Don't type the "period" at the end of the word, nor the quotes around the word either!) Notice that there is a short vertical line that blinks at the end of the letter "e" in Type. This line is called the **cursor**. Also note that the word Type appears in the formula toolbar line.

Because we are finished keying in the word "Type," we need to signal the Excel program that we wish to move on. There are several ways to do this. Perhaps the easiest way is simply to press the Enter or Return key on the keyboard. An alternative method is to move the mouse until the white arrow points at the green "check" mark to the left of the word Type in the formula toolbar line.

If you press the Enter or Return key, the highlighted cell moves from A1 to either A2 or to B1 and the word Type appears by itself in cell A1. If you point and click on the check mark on the formula toolbar line, the word Type appears in cell A1 and cell A1 remains highlighted (the bold, black line encircles the perimeter of the cell).

In either case, find the arrow keys on your keyboard; they appear as a set of four keys usually between the 26 letter keys on the left and the 10 number keys on the far right of your keyboard. Push several of them to see what happens. Using either the arrow keys or the mouse, make cell B1 the active cell on the spreadsheet.

Now key in the phrase "Number Sold" in cell B1 and then press the Enter key or point and click on the check mark on the formula toolbar line. (After we did this we made cell B2 the active cell so that we could clearly see all the words that we typed in cells A1 and B1). The following figure illustrates how your spreadsheet should look at this point. Notice that the phrase "Number Sold" has exceeded the right margin of cell B1 and some of the letters have spilled into cell C1.

TI-83 and Excel Basics

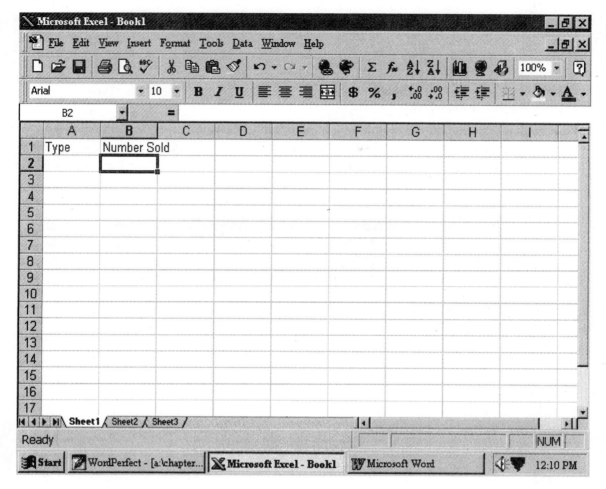

If we started entering data or text into cell C1 we will lose sight of the spillover from cell B1. What we would like to do is to increase the width of cell B1 so that the entire phrase "Number Sold" is completely contained inside cell B1. To widen cell B1 (and, in fact, the entire B column), move the mouse until it rests on the letter B in the column headings. The shape of the pointer should still be a white cross. Now, very slowly move the white cross to the right toward the letter C in the column headings. About halfway between the letters B and C the white cross changes shape. Stop the mouse when the white cross becomes the shape of a black vertical line with horizontal arrows pointing from the vertical line to the left and to the right.

Now click on the mouse and hold the button down; don't release the mouse button yet. Slowly slide the mouse to the right, toward the letter C. As you do this the right margin of the B column will increase. When you have moved far enough to the right past the end of the phrase "Number Sold," release the mouse button. Column B is now wider than the other columns and the entire phrase in cell B1 should be readable with no spillover to cell C1.

Now, go back to cell A1 and make it active. Move the pointer to the word "Type" in the formula toolbar line (right above the column letter C). Notice the pointer takes on yet another shape, this time resembling a large, capital I. Position the "I" pointer at the end of the word "Type" and click once. Now key in the rest of the phrase "Type of Bedding." Click on the green check mark when you are finished. Again the phrase exceeds the size of the cell. Repeat our previous instructions to increase the width of the A column so that the phrase "Type of Bedding" is completely visible in cell A1.

Make cell A2 active and key in the word "Single." Finish keying in the words Double, Queen, and King in cells A3, A4, and A5, respectively. Then move over to column B and key in the numbers 8, 12, 9, and 5 in cells B2, B3, B4, and B5, respectively. Your Excel spreadsheet should now resemble the following figure.

	A	B	C	D	E	F	G	H
1	Type of Bedding	Number Sold						
2	Single	8						
3	Double	12						
4	Queen	9						
5	King	5						
6								
7								

Saving Your Work

Now that we have entered our data set into an Excel spreadsheet, we need to "park" the data (or "save" the data) somewhere until we need it at a later date. We have two potential parking lots -- the hard drive built into our computer or a floppy disk that we can insert and remove from the disk drive.

Hard drives have a lot more parking space than floppy disks; hence, it is tempting to store all data sets on the hard drive. This is an acceptable storage spot, if you are working on your own computer. However, if you are working on a computer in a computer lab on campus, then you cannot use the hard drive to store your work. Let's use our current data set to demonstrate how to save a file using both parking lots.

To save the data set on your hard drive, move the white arrow to the word **File** on the main menu (the second horizontal line) and click once. Then scan through the menu that popped down and highlight the phrase **Save As**. Click on Save As and a new screen should appear over top of the spreadsheet. Refer to the following figure. At the beginning of the second horizontal line on this screen is the phrase "Save in:" and immediately following that phrase is the phrase "My Documents."

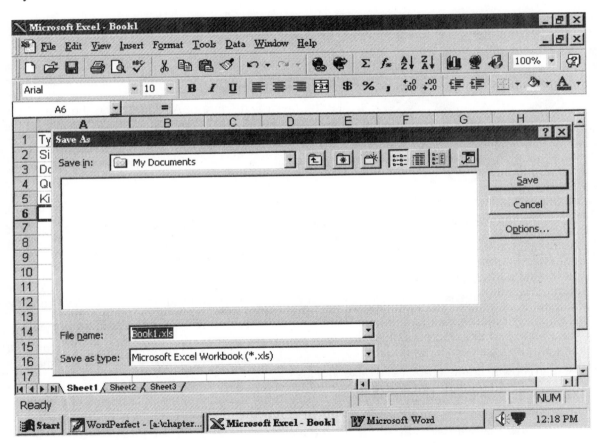

Look at the next-to-the-last line labeled "File name:" on this screen. Currently our file is named "Book1.xls." The last three letters -- xls -- are called an "extension" and signify that this file is an Excel file. (If we were typing and saving a letter using the Microsoft Word program, the extension of the file would have the letters "doc" instead of "xls." File extensions vary and are tied to the specific program that you are using.)

Instead of titling our file "Book1," let's name it "Bedding Sales." The phrase "Book1.xls" already should be highlighted, so we can immediately begin keying in the words "Bedding Sales." (If the phrase is not highlighted, rest the pointer on "Book1" and double click. This action highlights everything to the left of the period. Then begin keying in "Bedding Sales.")

You don't have to key in the extension ".xls." The Excel program will automatically add it to the file name if it isn't explicitly listed. Because the My Documents folder is located on your hard drive, clicking on the word **Save** near the upper right-hand corner of the screen will save the file on your hard drive.

To save your work on a floppy disk, insert a freshly formatted (blank) disk in the computer's disk drive and follow the previous instructions until your screen looks like the preceding figure. Again, change the title of the file from "Book1" to "Bedding Sales." (Follow the instructions listed in the previous paragraphs.)

After changing the title, look to the right of the "My Documents" phrase for the little triangle that is pointing down and click on it. Several options pop down, including one that says "3½ Floppy (A:)." Highlight the **3½ Floppy (A:)** option and click once.

Then click on **Save** in the upper right-hand corner of the screen. When the disk drive finishes making noises and the light goes out near the drive opening, remove the disk from the computer by pushing in the button that protrudes near the entrance to the disk drive.

We will reuse this data set so safely store the disk that you just retrieved from the disk drive. In Chapter 2, we will introduce more commands and buttons on the Excel spreadsheet as we learn how to create graphs and other images.

Ending Your Excel Session

Now that we have saved the file, we can close up shop. Click on the X symbol in the upper right-hand corner of the spreadsheet. The desktop screen should reappear. Then click on the Start button in the lower left-hand corner of the screen and highlight the "Shut Down ..." phrase. Click once and then follow the directions that appear on the screen.

PRETEST #1

Allow 20 Minutes
1 point each -- maximum score 10

1.1   Evaluate the following expression:

$$1.7 + \frac{.7}{25}(16 - 6) = ?$$

1.2   Convert .105 to a percent.

1.3   Convert 81.3% to a decimal.

1.4   Round-off 345.0149 to three decimal places.

1.5   Second quarter sales were $1.25 million. What is the percent change in sales, relative to first quarter sales of $.85 million? Round to one decimal place.

1.6   Use an inequality sign (< or >) to show which number is bigger, -1.23 or -1.32.

1.7 $|-7| = ?$

1.8   Rearrange the following numbers in order from low to high: 6, 0, -1, 1, -2, -3

1.9   Round the number 509,500 to the nearest thousand.

1.10   True or false: $|-5| < |3|$?

When we add, subtract, multiply, or divide, we are doing *arithmetic*. As elementary school students, we may have learned how to do these operations by hand, but today most of us rely on a hand-held calculator to do the number crunching for us. Unless we put a wrong number into the calculator, we are well-assured that the resulting number in the calculator display is correct. But when we have to perform a calculation that mixes the operations, we cannot be certain that our calculator yields the correct answer. We may have forgotten the correct sequence with which adding, subtracting, multiplying, or dividing should be performed or our calculator may not be programmed to perform the operations in the correct sequence. Let us start with a review of the correct ordering of arithmetic operations.

1.1   Order of Operations

Arithmetic operations, in general, conform to the following priority schedule:

RULES

PRIORITY OF ARITHMETIC OPERATIONS

1. *Exponentiation (raising a number to a power)*
2. *Multiplication or division*
3. *Addition or subtraction*

The priority schedule implies that when an expression involves all three operations, we should first concentrate our efforts on the exponents before we multiply or divide. Adding or subtracting comes last. For example, the expression $4 + 2 \cdot 3^2$ would be evaluated first by squaring 3 (to get 9); second, multiplying by 2 (to get 18); and third, adding 4 (to get 22):

$$4 + 2 \cdot 3^2 = 4 + 2 \cdot 9 = 4 + 18 = 22.$$

The only exception to the arithmetical operation priority schedule occurs when parentheses are present in expressions. The operation within the parentheses then takes precedence over any other operation. For example, consider the following:

$$(1 - 5)^2 + (4 - 5)^2 + (10 - 5)^2$$

In this case we will work within the parentheses first and perform the indicated subtractions, then we will square, and finally we will add the squares:

$$\begin{aligned}(1 - 5)^2 + (4 - 5)^2 + (10 - 5)^2 \ &= (-4)^2 + (-1)^2 + (5)^2 \\ &= 16 + 1 + 25 \\ &= 42\end{aligned}$$

One "trick" of subtraction that we sometimes forget occurs when we subtract a negative number, such as subtracting -7 from 24:

$$24 - (-7) = 24 + 7 = 31$$

The double minus signs turn into a positive sign and we end up adding.

Our need to understand the correct sequence of operations will become apparent when we study descriptive statistics. For instance, when we are finding the median for a grouped frequency distribution, we might encounter a computation like this:

$$1.10 + \frac{.92}{8} (10 - 4)$$

The correct answer of 1.79 is found first by performing the subtraction inside the parentheses to get 6; second, multiplying .92 by 6 and then dividing by 8 to get .69; and third, adding 1.10 to .69.

SUMMARY EXAMPLE
Order of Operations

Evaluate $83.2 + \dfrac{8.4}{54} (34 - 25)$

Do this first:	Work within parentheses	$(34 - 25) = 9$
Do this second:	Multiply/divide	$\dfrac{8.4}{54} (9) = 1.4$
Do this third:	Add	$83.2 + 1.4 = \mathbf{84.6} \leftarrow$ **Answer**

Does this statement sound familiar: "I know the rules; I just don't know how to *apply* them to a word problem." Word problems (or story problems, as some people call them) wrap a short story around a specific math skill. Your job is twofold -- understand the story and then decipher which rule or formula applies to the problem. While not all word problems are solved in the same manner, there are some actions you can take to help you get started. Below are recommended problem solving diagnostics for word problems. Each may not apply to every problem.

PROBLEM SOLVING DIAGNOSTICS
FOR WORD PROBLEMS

1. Underline or highlight key words in the story.
2. Circle all numbers given in the problem and try to attach a symbol or meaning to each number. (Note that some numbers may be meaningless and can be ignored.)
3. Draw a rough picture that captures the image of what is being described, if applicable.
4. Find the sentence with the question mark (?) at the end of it. Read it over and over until you can write down exactly what quantity, symbol, or term you are to find.
5. Write down the appropriate formula and try to apply it, if applicable. If there is no appropriate formula, start making some preliminary calculations. Each time your calculations yield a result, ask yourself "What does this number represent?"

WORD PROBLEM EXAMPLE

As an example, consider the following word problem:

April and Christy share a two bedroom apartment on 3rd Street. A series of recent storms caused the creek behind their apartment to flood, damaging the carpet in Christy's bedroom. The bedroom measures 10 feet wide by 12 feet long, plus a closet that measures 3 feet deep by 5 feet across. Assuming no waste, how many square yards of new carpet do they need to buy?

Let's go through the Problem Solving Diagnostics step-by-step:

1. **Key words:**   Bedroom, closet, square yards
2. **Numbers:**   Two bedroom (meaningless)
    3rd Street (meaningless)
    10 feet (width of bedroom)
    12 feet (length of bedroom)
    3 feet (width of closet)
    5 feet (length of closet)

3.   **Picture:**

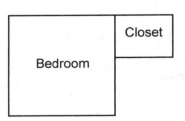

4.   **Question mark:**   Find the area of the bedroom plus the closet floor (in square yards).

5.   **Computations:**

Area   = length x width

       = (12 X 10) + (5 X 3)

              area of bedroom   area of closet

       = 120 + 15 = 135 square feet

Square yards = Square feet/9

             = 135/9

             = 15 square yards

Notice in step #5 that we *applied* our knowledge of the priority of arithmetic operations: we multiplied first and then added. Also note that we had to convert square feet to square yards, so we had to know that 9 square feet = 1 square yard.

1.2   Decimals, Percents, and Percent Change

A *decimal fraction* is a number between 0 and 1 representing a ratio of two numbers in which the denominator must be 10, 100, 1000, 10000 and so on. For example, .25, .1, and .873 are *decimal fractions* since they can be expressed as the following ratios:

$$.25 = \frac{25}{100}, \quad .1 = \frac{1}{10}, \text{ and } \quad .873 = \frac{873}{1000}$$

In conversation, the term "decimal fraction" is sometimes shortened to the term *decimal*.

Often, we wish to change these decimals into whole numbers (and fractions thereof) by converting them to percents. A *percent* is a number between 0 and 100 representing the numerator of a decimal fraction in which the denominator must be 100. Thus, 25%, 10%, and 87.3% are percents since they are the numerators of the following decimals:

$$\frac{25}{100}, \frac{10}{100}, \text{ and } \frac{87.3}{100}$$

Our interest in percents and decimals is concerned with being able to obtain one from the other. To convert a decimal into a percent, move the decimal point two places to the right and attach the percent symbol (%).

SUMMARY EXAMPLE
Decimal to Percent

	Convert .049 to a percent	
Do this first:	Move decimal point two places to the **right**	.049 becomes 04.9
Do this second:	Attach % symbol	**4.9% ← Answer**

Conversely, to create a decimal from a percent, drop the percent sign and move the decimal two places to the left as indicated in the example.

SUMMARY EXAMPLE
Percent to Decimal

	Convert .91% to a decimal	
Do this first:	Drop % symbol	.91% becomes .91
Do this second:	Move decimal point two places to the **left**	.91 becomes **.0091 ← Answer**

A common use of percents is in reporting a percent change from one time period to the next. A *percent change* is the percent difference in two values relative to one of the values. A percent change differs from a percent in that a percent change is <u>not</u> restricted to being a number between 0 and 100. For example, a percent change could be –15% or 200%. In general, to find a percent change we use the "old" value as our point of reference. Here is the basic formula for finding a percent change:

Arithmetic

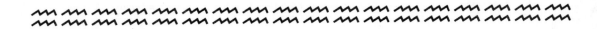

«*FORMULA*»
PERCENT CHANGE RELATIVE TO OLD VALUE

$$\text{Percent Change} = \frac{\text{New value} - \text{Old value}}{\text{Old value}} \times 100\%$$

For instance, suppose during the first quarter of last year an automaker sold 25,000 vehicles. During the first quarter of this year, the automaker sold 24,000 vehicles. Then the percent change in sales relative to the first quarter of last year is:

$$\text{Percent change} = \frac{24,000 - 25,000}{25,000} \times 100\%$$

$$= \frac{-1,000}{25,000} \times 100\% = -.04 \times 100\% = -4\%$$

Sales were down 4 percent from the same period in the previous year.

SUMMARY EXAMPLE
Percent Change

Attendance at last year's show was 16,000 people.
This year 16,800 people attended the show. What was the percent
change in attendance, relative to last year's show?

Do this first:	Subtract: New – Old	16,800 – 16,000 = 800
Do this second:	Divide: $\dfrac{\text{New} - \text{Old}}{\text{Old}}$	$\dfrac{800}{16,000} = .05$
Do this third:	Multiply: $\dfrac{\text{New} - \text{Old}}{\text{Old}} \times 100\%$	.05 X 100% = **5%** **Answer** ↗ Attendance increased 5 percent over last year's figure.

We will encounter the concept of percent change in the study of time series. For time series applications we will be interested in the percent error in a forecast relative to the actual value. The formula for percent error is:

«FORMULA»

PERCENT ERROR RELATIVE TO ACTUAL VALUE

$$\text{Percent error} = \frac{\text{Actual value} - \text{Forecasted value}}{\text{Actual value}} \times 100\%$$

If, for example, the actual value were 8 and the forecasted value were 7.4 then the percent error in the forecast would be:

$$\text{Percent error} = \frac{8 - 7.4}{8} \times 100\% = \frac{.6}{8} \times 100\% = .075 \times 100\% = 7.5\%$$

The actual value was 7.5% higher than the forecasted value.

1.3   Rounding

As we perform computations that involve multiplication, division, and raising numbers to non-integer powers, we may produce decimal fractions with many digits. Although rounding may be necessary, we caution against rounding intermediate calculations. Round only the final answer, as indicated in our round-off rule for decimals below.

RULES
ROUNDING OFF DECIMALS

1.   *Determine the number of decimal places desired to the right of the decimal.*
2.   *Examine the digit in the succeeding decimal place.*
   a.   *If the digit is 5 or more, increase the preceding digit by 1 and eliminate all succeeding digits.*
   b.   *If the digit is 4 or less, leave the preceding digit as is and eliminate all succeeding digits.*

Let us demonstrate this rule using the number 7.149501. If we wished to round-off to one decimal place, we would examine the digit 4 in the second decimal place and, according to part 2b above, round the number to 7.1. To round to two decimal places, the digit 9 in the succeeding decimal place forces us to increase the 4 in the second decimal place by one, so that the rounded value is 7.15. Rounding to three decimal places yields 7.150: the digit 5 in the fourth decimal place causes us to increase the 9 in the third position by one, which, in turn, increases the 4 in the second position by one.

One variation of our rounding procedure -- *rounding up* -- will occur occasionally in statistics. The round-up rule for decimals is stated next.

RULES
ROUNDING UP DECIMALS

1. Determine the number of decimal places desired to the right of the decimal.
2. Regardless of the digit in the succeeding decimal place, increase the preceding digit by one and eliminate all succeeding digits.

For instance, to round-up the number 8.1402 to two decimal places, we would increase the 4 to a 5, eliminate all succeeding digits, and report 8.15; to round-up to one decimal place, 8.2; and to round-up to zero decimal places (that is, to round-up to a whole number), 9.

The principle of rounding decimal fractions can be extended to rounding to whole numbers as well. This time we focus on decimal places to the *left* of the decimal point instead of to the right. Rounding to whole numbers means that <u>no</u> numbers will appear to the right of the decimal point. Previously we rounded according to the desired <u>number</u> of decimal places to the right. However, in rounding to whole numbers we usually refer to the desired <u>accuracy</u> of the rounded numbers rather than specifying the number of decimal places. The following chart lists some common accuracies and examples of correspondingly rounded numbers. Note that all rounded numbers end in a zero or a sequence of zeroes, depending on the accuracy.

Rounding Accuracy To the nearest ...	Number of 0s in ending sequence of rounded number*	Example of rounded whole number
ten	1	10 or 740 or 4,000
hundred	2	300 or 6,800
thousand	3	2,000 or 85,000
hundred thousand	5	400,000 or 17,900,000
million	6	16,000,000

* The number of ending 0s could *exceed* this figure. For example the number 4,000 could be a rounded number to the nearest ten.

Rounding to whole numbers depends on two things: the desired accuracy of the rounded number and the *trailing accuracy digit* in the original number. By *trailing accuracy digit* we mean the immediately succeeding digit, once accuracy has been specified. For example, if the accuracy is

Arithmetic

hundreds, then the trailing accuracy digit is the digit in the tens position. If the trailing accuracy digit is 5 or more, then we increase the preceding digit by one and add the appropriate number of 0s according to the previous table. If the trailing accuracy digit is 4 or less, we leave the preceding digit as is and add the appropriate number of 0s.

As an example, consider the number 32,068,915. Let us round this number to several different accuracies, as indicated below:

Round to the nearest ...	Trailing accuracy digit in 32,068,915	Rounded number	Reason
ten	5	32,068,920	Trailing accuracy digit is 5 or more; increase preceding digit (1) to 2 and add one zero.
hundred	1	32,068,900	Trailing accuracy digit is 4 or less; leave preceding digit (9) as is and add two zeroes.
thousand	9	32,069,000	Trailing accuracy digit is 5 or more; increase preceding digit (8) to 9 and add three zeroes.
hundred thousand	6	32,100,000	Trailing accuracy digit is 5 or more; increase preceding digit (0) to 1 and add five zeroes.
million	0	32,000,000	Trailing accuracy digit is 4 or less; leave preceding digit (2) as is and add six zeroes.

In statistics you will be using your calculator often to solve multi-step problems. Rounding off intermediate answers can lead to terribly incorrect final answers. Do not round-off until the last calculation has been completed. Calculators can help you avoid problems if you learn how to use the memory features of your calculator (assuming your calculator has memory capabilities). This will make it unnecessary to write down intermediate results and will improve the accuracy in the final answer. If you think accuracy isn't important, let us remind you that an error of less than $1/50$th the width of a human hair in a lens in the Hubble telescope cost several million dollars to repair, as well as many lost opportunities to observe space phenomena.

When the word *round* is used in an exercise, you should *round-off* (not up). For example, the question "What is the percent error? Round to one decimal place." means we should round-off the percent error to one decimal place.

SUMMARY EXAMPLE
Rounding

Round 96.3516 to	... 1 decimal place	... 2 decimal places	... the nearest ten

 ... TO 1 DECIMAL PLACE:

Do this first:	Look at digit in 2nd decimal place: 96.3**5**16
Do this second:	Because the 2nd digit is 5 or more, increase the preceeding digit from 3 to 4: **96.4 ← Answer**

 ... TO 2 DECIMAL PLACES:

Do this first:	Look at digit in 3rd decimal place: 96.35**1**6
Do this second:	Because the 3rd digit is less than 5, leave the preceeding digit 5 as is: **96.35 ← Answer**

 ... TO THE NEAREST TEN:

Do this first:	Look at trailing accuracy digit: 9**6**.3516
Do this second:	Because the trailing accuracy digit is 5 or more, increase the preceeding digit from 9 to 10 and add one zero: **100 ← Answer**

Arithmetic

▷▶ USING YOUR TI-83 ▷▶

The TI-83 can do rounding, but only to a given number of decimal places. It cannot round to the nearest ten, hundred, etc.

〰〰〰〰〰〰〰〰〰〰〰〰〰〰〰〰〰〰〰〰〰〰〰〰〰〰

USE THE TI-83   **to round 96.3516 to**   **...3 decimal places**
                                                ...0 decimal places (a whole number)

...to 3 decimal places   ▷▶   *Press MATH, right arrow, 2, 96.3516, comma (above the 7), 3, ENTER.*

DISPLAY: **96.352**

...to 0 decimal places   ▷▶   *Same as above, but after the comma key, replace "3" with "0."*

DISPLAY: **96**

〰〰〰〰〰〰〰〰〰〰〰〰〰〰〰〰〰〰〰〰〰〰〰〰〰〰

1.4   Inequalities

In an area of statistics called hypothesis testing we have to judge which of two numbers is larger (or smaller). On the surface this seems to be a simple task; for instance, 8 is clearly larger than 5. One way to write this fact using an inequality sign is 8 > 5. Another way is to write 5 < 8. Notice that the smaller end of the symbols < and > points to the smaller number (5 in the preceding examples). The *greater than* (>) and *less than* (<) symbols are used to indicate the ordering of numbers along a number line. Smaller numbers are to the left of larger numbers: smaller number < larger number. Larger numbers are to the right of smaller numbers: larger number > smaller number.

If both numbers are positive, then it is a simple matter to order them: subtract the two numbers, as in a – b. If the result is a positive number, then a > b. If the result is a negative number, then a < b. For example, to establish the relationship between 6.9 and 6.95, we subtract:

$$6.9 - 6.95 = -.05$$

Because the result is a negative number, we write 6.9 < 6.95.

If one number is positive and the other negative, then the positive number is always greater than the negative number (or, equivalently, the negative number is always less than the positive number). For example, the relationship between the numbers –3.1 and 0.2 is:

$$-3.1 < 0.2 \text{ (or } 0.2 > -3.1)$$

What if both numbers are negative? For example, suppose a weather forecaster predicted that the temperature was going to *fall* from –5°F to –20°F. We think you would agree that –20°F is colder than –5°F, so naturally –20 < –5.

Even for two negative numbers, a and b, our previous rule still works: Subtract a – b; if the result is a positive number, then a > b. If the result is a negative number, then a < b. For the two temperatures –20°F and –5°F, we subtract (don't forget what we said earlier in the chapter that subtracting a negative number is actually adding):

$$-20 - (-5) = -20 + 5 = -15$$

Because the result is a negative number, we write:

$$-20 < -5$$

Establishing the order between two numbers like –20 and –5 can be extended to the problem of *ordering a set of numbers*. In descriptive statistics, we need this ordering ability when we are determining the *median* of a set of data. Identifying the median requires us to first arrange the data in sequence from low numbers to high numbers (or vice versa) by listing the smallest number first, then the second smallest number, and so on up to the largest number.

For example, suppose we wished to order the following data set from low to high: 2, 5, –3, 8, 7, 3, –4. We start with the smallest number, which is –4 (because –4 < –3). The second smallest number is –3. From there we identify 2 as the next smallest, and we continue until we arrive at the largest number, 8. The complete ordering from low to high is:

$$-4, -3, 2, 3, 5, 7, 8$$

because this sequence of numbers conforms to the following string of inequalities:

$$-4 < -3 < 2 < 3 < 5 < 7 < 8$$

 **USING YOUR TI-83** ▷▶

The TI-83 can "test" values to let you know the direction of an inequality.

USE THE TI-83	to see if	...1.35 > 1.347	...-4.12 > -4.1

Is 1.35 > 1.347?   ▷▶   *Press 1.35, 2nd, MATH, 3, (to choose "greater than"), 1.347, ENTER.*

DISPLAY: **1, which denotes "true"**

Is -4.12 > -4.1?   ▷▶   *Press (-), 4.12, 2nd, MATH, 3, (-),4.1, ENTER.*

DISPLAY: **0, denoting that -4.12 > -4.1 is a false statement. (Note that you could do the same tests with the "less than" symbol instead of the "greater than" symbol if you wished. This would reverse the 0 or 1 displays shown above, however.)**

1.5 Absolute Value

The *absolute value* of a positive or negative number strips away the sign and leaves just the numerical part of the number standing. For example, the absolute value of +12 is 12; the absolute value of -8 is 8; and the absolute value of 0 is 0. We use vertical lines around the number to symbolize absolute value:

$$|+12| = 12 \qquad\qquad |-8| = 8 \qquad\qquad |0| = 0$$

More often than not, positive numbers are written without the + sign;   thus, instead of writing $|+12|$ = 12, you will frequently see $|12|$ = 12.

One use of absolute values (and inequalities) in statistics occurs in hypothesis testing when we form decision rules (or rejection regions). Although you may be unfamiliar with these statistical terms, here is an example of a decision rule: *If $|z| > 1.96$, take a certain action.*

The key is to decide whether the inequality $|z| > 1.96$ is true or false based on a computed value of z. For instance, if z = -1.83, the inequality $|z| > 1.96$ is false because $|z| = |-1.83| = 1.83$, and 1.83 is <u>not</u> > 1.96.

Chapter 1 Exercises

1.1 $(1 + .017) (1 - .024) (1 + .105) = ?$

1.2 $7.0 + \dfrac{16.4}{46} (86 - 17) = ?$

1.3 $.191 + \dfrac{.035}{120} (50 - 38) = ?$

1.4 $[-1-(-4)]^2 + [-5-(-4)]^2 + [-6-(-4)]^2 + [-6-(-4)]^2 + [-2-(-4)]^2 = ?$

1.5 $(2-3)^2 + (0-3)^2 + (6-3)^2 + (3-3)^2 + (9-3)^2 + (2-3)^2 + (0-3)^2 + (2-3)^2 = ?$

1.6 $(2-3)^3 + (0-3)^3 + (6-3)^3 + (3-3)^3 + (9-3)^3 + (2-3)^3 + (0-3)^3 + (2-3)^3 = ?$

1.7 $4^2 + 6^2 + 1^2 + 0^2 + 2^2 - \dfrac{(4 + 6 + 1 + 0 + 2)^2}{5} = ?$

1.8 Convert the following percents to decimals:

    (a)   −2.70%
    (b)   −4.11%
    (c)   13.09%
    (d)   1.84%
    (e)   −.79%

1.9 Convert the following decimals to percents:

    (a)   .057
    (b)   .180
    (c)   .7011
    (d)   −.216
    (e)   .005

1.10 Round-up the number 1.09615 to:

    (a)   4 decimal places
    (b)   3 decimal places
    (c)   2 decimal places

continued on the next page ...

Arithmetic

Chapter 1 Exercises (continued)

1.11   Round-off the following to two decimal places. To three decimal places.

    (a)   3.296501
    (b)   11.028133
    (c)   −5.099627
    (d)   2.239366
    (e)   2.265893
    (f)   0.578000

1.12   A company processed 925 employment applications in its first year of operation and 625 the following year. What was the percent change in applications, relative to the first year? Round to 2 decimal places.

1.13   The initial share price of a new stock offering was $8. One week later the share price was $20. One month later, the share price was $6. Find the percent change in share price, relative to the initial share price...

    (a)   One week later
    (b)   One month later

1.14   Find the percent error in the following forecasts. Round to 2 decimal places.

    (a)   Actual value = 62.1; forecasted value = 65
    (b)   Actual value = 13.75; forecasted value = 12.21
    (c)   Actual value = 7; forecasted value = 7
    (d)   Actual value = .8; forecasted value = 2.

1.15   Evaluate the following:

    (a) $|-35|$
    (b) $|+1.73|$
    (c) $|\,0.05|$

1.16   Establish the ordering relationship (> or <) between the following pairs of numbers:

    (a)   1.35 ? 1.347
    (b)   −1.28 ? −1.34
    (c)   −2.02 ? −1.65
    (d)   −7.7 ? .01
    (e)   6.1 ? 6.1001
    (f)   −4.9 ? −4.8

continued on the next page ...

Chapter 1 Exercises (continued)

1.17 The local electrical utility has a two-tiered rate structure for residential homes depending on the amount of electricity consumed (as measured in kilowatt hours, KWH). For consumption of less than 1000 KWH per month, the rate is 5.2 cents per KWH plus a base fee of $10. The rate for those who use more than 1000 KWH per month is 5.2 cents per KWH for the first 1000 KWH, then 4.8 cents per KWH for each KWH exceeding 1000, plus a base fee of $10. How much would the electric bill be for the following consumption totals?

    (a)   810 KWH
    (b)   1485 KWH

1.18 Tom is watching his daily intake of fat. For lunch he had a hamburger, an order of french fries, and a medium soft drink at a nearby fast food restaurant. Suppose there are 342 calories in the hamburger, 400 calories in the french fries, and 75 calories in the soft drink. The percent of fat calories is 50% in the hamburger, 45% for the french fries, and 0% for the soft drink. How many grams of fat did Tom ingest for lunch? (Assume that there are 9 calories in each gram of fat.)

1.19 Rearrange the following numbers in order from low to high:

8.50, 8.25, 8.37, 8.38

1.20 Rearrange the following numbers in order from low to high:

-4, -6, -1, 0, -2

1.21 Round the number 75,546,532 to the nearest

    (a)   ten
    (b)   hundred
    (c)   thousand
    (d)   hundred thousand
    (e)   million

1.22 Round the number 901,655 to the nearest

    (a)   ten
    (b)   hundred
    (c)   thousand
    (d)   hundred thousand
    (e)   million

POSTTEST #1
Allow 25 Minutes
1 point each -- maximum score 10

1.1 Evaluate the following expression:

$$5^2 + 3^2 + 7^2 - \frac{(5 + 3 + 7)^2}{3} = ?$$

1.2 Convert .124 to a percent.

1.3 Convert .8% to a decimal.

1.4 Round-off 42.385 to two decimal places.

1.5 Market share changed from 10.4% last year to 10.8% this year. What is the percent change in market share this year, relative to last year? Round to one decimal place.

1.6 Use an inequality sign (< or >) to show which number is bigger, −9 or −9.1.

1.7 $|-62| = ?$

1.8 Jennifer returned from her trip to Europe with 13 British pounds, 77 French francs, and 20 German marks. If the current exchange rates are as follows,

Foreign currency	Currency per U.S. Dollar
British pound	.65
French franc	5.50
German mark	1.60

how much money, in U.S. dollars, did Jennifer have?

1.9 Rearrange the following numbers in order from low to high:

$$-7, -3, -4, -1, -5, 0$$

1.10 Round-off 9,050 to the nearest hundred.

CHAPTER TWO
Graphs

PRETEST #2 - Allow 10 Minutes
Point Values -- maximum score 7
3 points for 2.1, 1 point each for 2.2 - 2.5

2.1 Identify three omissions in the following graph:

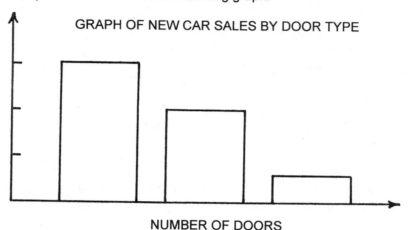

GRAPH OF NEW CAR SALES BY DOOR TYPE

NUMBER OF DOORS

2.2 Refer to Question 2.1. Name the type of graph illustrated.

2.3 A sample of 80 stocks revealed the following exchanges on which the stocks are listed.

Exchange	Number
New York Stock Exchange (NYSE)	42
NASDAQ	30
American Stock Exchange	8
	80

If we were drawing a pie chart to represent these data, what percent of the pie would be taken up by the NYSE slice?

continued on the next page...

2.4 Refer to Question 2.3. What angle would be needed to represent the NASDAQ category?

2.5 Which of the following graphs depicts the line Y = .5X + 2?

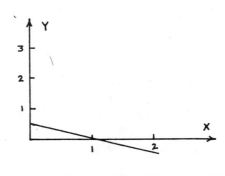

a.   b.   c.   d.

Although the development of computer graphics has almost eliminated the need to construct graphs by hand, we believe a short review of graphics will accomplish two purposes. First, it will enable us to recognize good (and bad) graphs constructed by a computer software routine (or by hand). Second, should we be required to produce a graph, we will be able to refer to this section for reference.

In the study of statistics, we will be exposed to three main varieties of graphs: bar graphs, pie charts, and bivariate graphs. Within these broad categories specialized graphs that represent variations of the main types exist. For example, a bar graph may be called a *histogram*, depending on what is plotted on the horizontal axis. Also, a bivariate graph becomes a *scatter plot* if we do <u>not</u> connect the points with line segments.

Before demonstrating the various graphical techniques, we must mention several guidelines to follow in preparing a graph.

RULES
CONSTRUCTING GRAPHS

1. *Always use graph paper, preferably with a light blue grid that will not reproduce when photocopied.*
2. *Never draw a freehand line; always use a straight edge.*
3. *Always label the axes (or slices in a pie chart) and provide a title for the graph.*

The key here is that we encourage you to make every graph appear as professional as possible. If you never take the time to draw a graph accurately for a homework assignment, then when the time comes to draw one for a presentation in your first job you will have no prior experience.

2.1 Bar Graphs

The term *bar graph* suggests that the graph is made up of a sequence of bars or rectangles. To illustrate its construction, suppose a furniture store sold the following numbers of bedding sets during a holiday weekend sale.

Type of Bedding	Number Sold
Single	8
Double	12
Queen	9
King	5

To graph these data into a bar graph, we would set up two axes: a horizontal axis to represent the variable *type of bedding* and a vertical axis to represent the *number sold*. In drawing these axes we recommend that the horizontal axis be longer than the vertical axis is tall. As a general rule, the height of the vertical axis should be $^2/_3$ to $^3/_4$ the length of the horizontal axis.

Next, along the horizontal axis we allocate space for the four categories of bedding. We scale the vertical axis from 0 through 14, say. Generally, the scale should be slightly larger than the largest number, 12 in this case, to be plotted. Above each category on the horizontal axis, we erect a rectangle with a height corresponding to the number of bedding sets of that type sold. Make each rectangle the same width and allow a gap between the rectangles rather than drawing them so that they touch. Finally, label each axis and title the graph. The following is an example of the finished product.

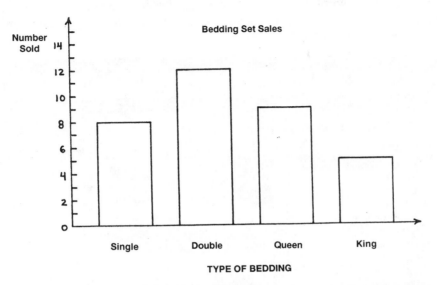

When the number of categories is large, we may write the frequency corresponding to each category at the top of the respective rectangle. This helps the viewer who cannot visually trace the height of each rectangle back to the vertical axis to determine the frequency.

Bar graphs also can be drawn "sideways," with the categories on the vertical axis and the numbers on the horizontal axis. Another option is to rearrange the categories on the horizontal axis so that the tallest rectangle is first, followed by the second tallest, and so on down to the shortest rectangle at the far right.

 **USING EXCEL**

There are numerous graphing options with Excel using a special key called the **Chart Wizard** key. This key is found on the standard toolbar, which usually appears as the third line down from the top of the screen. Before proceeding, you should verify that the standard toolbar is displayed on your screen.

Click on View in the main menu (second line from the top) and then highlight the word Toolbars. A secondary menu listing at least a dozen other toolbars should appear, and one of the options on this list should be the word Standard. If a box with a checkmark appears beside the word

Standard, then this toolbar will appear on your screen. If there is no checkmark beside the word, merely point and click on Standard and a checkmark will appear.

Scan across this toolbar until you see an icon that resembles a graph depicting a horizontal and a vertical axis with three colorful columns extending up from the horizontal axis. Move your mouse until the arrow rests on this icon for a few seconds. If you are pointing at the correct icon, Excel will display the name **Chart Wizard**.

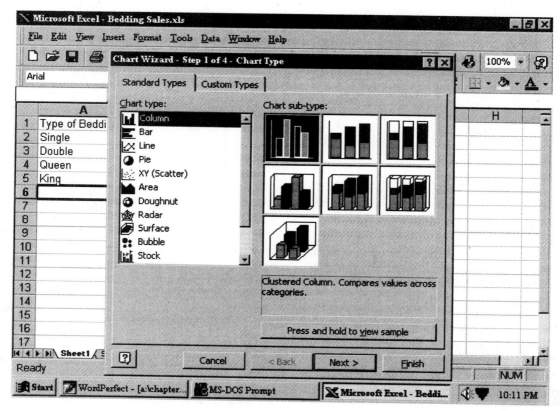

In Chapter 0, we practiced entering data into an Excel spreadsheet and saving the data in a file. In particular, the data set that we used at that time was the same as the bedding data found at the beginning of this section. We called the data set "Bedding Sales." Either retrieve this file now or key in the data in columns A and B. Row 1 should contain the labels "Type of Bedding" and "Number Sold" in columns A and B, respectively, while rows 2 through 5 should contain the data.

Point and click on the Chart Wizard key; your screen should look like preceding figure. Examine this figure closely. Across the top of the Chart Wizard window is the notation "Step 1 of 4 - Chart Type." This means that there will be three more screens to follow this one. When we finish with this screen, we will eventually point and click on the word "Next" near the lower right hand corner of the Chart Wizard window.

First, however, we must select the type of chart that we wish to draw. Notice that many chart types are listed, including ones called "Column" and "Bar." The icons to the left of these words indicate the difference in the two charts: Column charts display the rectangles up-and-down, while Bar charts display the rectangles sideways.

We wish to create a graph with the bars appearing up-and-down, so we will select the Column type. (Note: The vocabulary used in Excel is more precise than that used in the field of statistics. Most statistics texts refer to graphs with separated rectangles as "bar" graphs, regardless of the direction of the rectangles. Excel distinguishes the graphs based on direction. So, statistically speaking, we are creating a "bar" graph, which is actually a "column" graph in the Excel lexicon.)

Immediately to the right of the chart types in the Chart Wizard window is a series of different styles of column graphs, called "Chart sub-type:." Below these figures is a bar with the phrase "Press and hold to view sample." Point and click at one of the chart sub-types, and then point, click, and hold on the "Press and hold..." bar. Excel immediately displays a preview of our bedding data in the format of the selected chart sub-type.

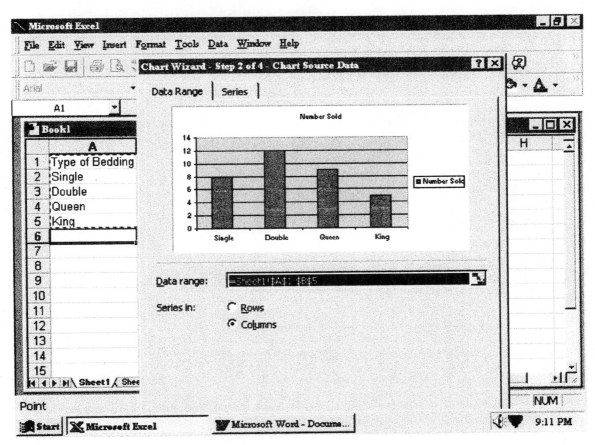

Select the one called "Clustered Column" (the one in the upper left hand corner in the Chart sub-types) and then point and click on the word Next. Step 2 in the Chart Wizard window appears, as shown on the previous page. In this step we must identify the source of the data for constructing the graph. If we have retrieved only the data set for which we wish to create the graph, then Excel automatically recognizes the appropriate rows and columns and proceeds to draw the graph. Otherwise, we have to specify in the "Data range" box where Excel will find the data.

Click on Next to go to Step 3 - Chart Options, where we can edit the title or axis labels. See the figure below. Currently the title of the graph is "Number Sold," as seen in the Chart title box. Let's change the title to "Bedding Set Sales." Press the **Tab** key which highlights the phrase "Number Sold" in black, and begin typing the new title, Bedding Set Sales.

Again press the **Tab** key to activate the "Category (X) axis:" box. In this box key-in the phrase "Type of bedding." This phrase will immediately appear beneath the categories on the horizontal axis.

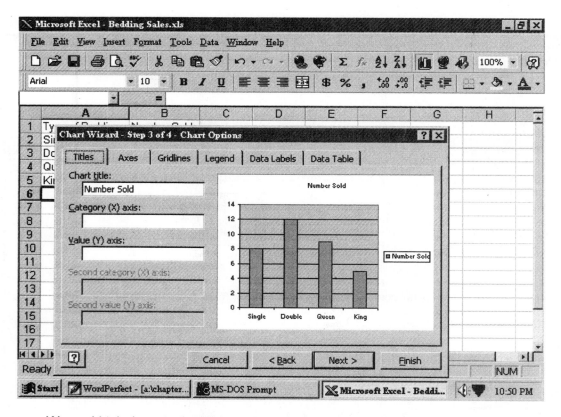

We could tab down to the "Value (Y) axis:" and key-in a label for the vertical axis (such as "Number sold"), but the graph already alerts the reader to this information with the little box showing this legend to the right of the graph. We don't need to repeat it. If your screen resembles the following figure, then click on Next.

Graphs

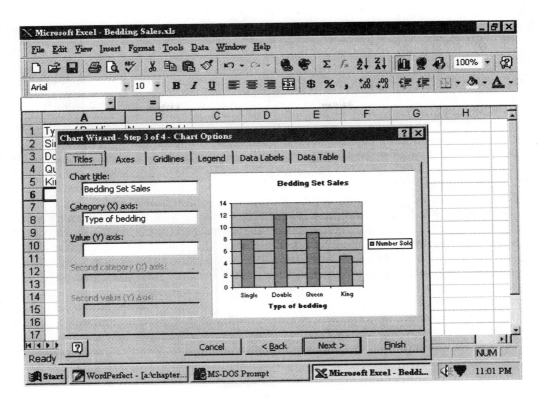

In step 4 of the Chart Wizard window we must decide where we wish to paste our newly created graph. We have two options, as a new sheet or as an object in our current spreadsheet. If not already selected, chose the latter option and then point and click on the word Finish in the lower right hand corner. Your graph should now resemble the figure to the right, which will appear in your Excel spreadsheet.

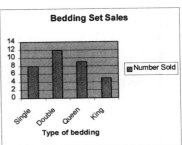

EXCEL F.Y.I.

To create a bar graph...

    (1) ... retrieve an existing data set from a file (or key-in a new data set),

    (2) click on the **Chart Wizard** key,

    (3) select Column from the Chart types,

    (4) select Clustered Column from the Chart sub-types; click Next,

    (5) highlight the rows and columns in the spreadsheet where the data reside
        (Note: this may already be done); click Next,

    (6) key-in appropriate chart title and label for X-axis; click Next,

    (7) place chart as an object in your current spreadsheet; click Finish.

SUMMARY EXAMPLE
Bar Graph

Draw a bar graph for these data:

Favorite pizza topping	Number of respondents
Pepperoni	45
Ground beef	80
Mushrooms	25
Other	20

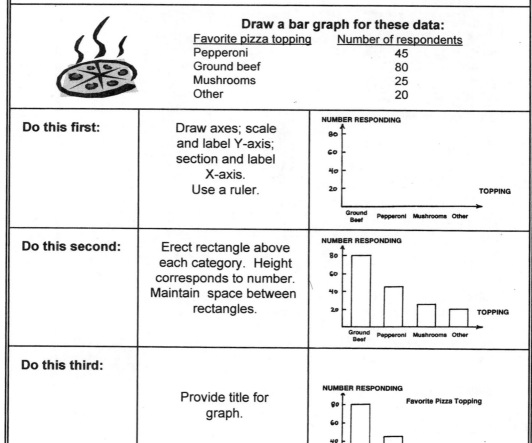

Do this first:	Draw axes; scale and label Y-axis; section and label X-axis. Use a ruler.
Do this second:	Erect rectangle above each category. Height corresponds to number. Maintain space between rectangles.
Do this third:	Provide title for graph.

When the variable to be plotted on the horizontal axis consists of numbers representing measurements or counts, rather than categories, we usually draw the rectangles so that they are adjacent, or touching each other, and call the resulting graph a *histogram*. For instance, suppose we needed to graph a set of data representing the length of time for a sample of 35 long distance phone calls. Further suppose the data have been organized into the following table:

Length of Phone Call (in seconds)	Number
0 -- under 120	11
120 -- under 240	13
240 -- under 360	5
360 -- under 480	4
480 -- under 600	2

In most statistics textbooks the term *frequency*, denoted by the symbol *f*, is used to signify the column titled "Number" in the above table.

Again, we construct a vertical axis for the *number* of phone calls and a horizontal axis for the *length of phone call* such that the vertical axis is about $^2/_3$ to $^3/_4$ as tall as the horizontal axis is long. Then, we scale each axis: the vertical from 0 through 15, and the horizontal in increments of 120. So that our first rectangle, corresponding to the 0 to 120 time interval, does not interfere with the vertical axis, we will shift the value of 0 on the horizontal axis a short distance to the right. Normally, the point of intersection of the horizontal and vertical axes is understood to be the zero-value for both scales. To show the viewer that we have altered this convention we will indicate a "break" in the horizontal axis with a special V-shaped symbol (see figure below). (Though not shown, another popular break symbol is a pair of parallel "squiggles," $\int\int$.)

Next, draw a rectangle above each time interval, making the height correspond to the number of phone calls. Each new rectangle should begin on the line where the last one ended. (Do not leave a space between rectangles, unless a particular interval has a count of 0.) Finally, label the axes and provide a title, as indicated below.

Duration of a Sample of Long Distance Phone Calls

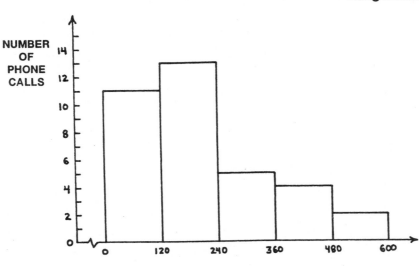

LENGTH OF PHONE CALL, in seconds

A graph of a *discrete probability distribution* is a type of histogram in which we plot the individual values of a variable on the horizontal axis and the associated probability on the vertical axis. Let X represent the horizontal axis and P(X=x) represent the vertical axis. The notation P(X=x) means "the probability that the variable X has the value x." A discrete probability distribution looks like the following graph.

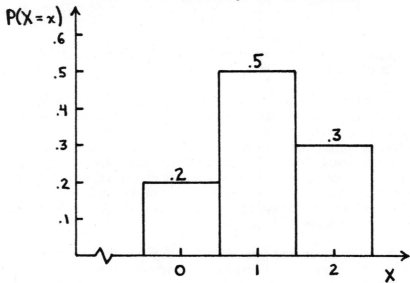

Discrete Probability Distribution for X

Notice that the rectangles are touching and that each rectangle is <u>centered</u> over the specific value. In some statistics textbooks, discrete probability distributions may appear as a series of vertical lines or spikes above each X-value instead of as a series of rectangles.

▷▶USING YOUR TI-83 ▷▶

The TI-83 can graph a histogram, although the axes will be not labeled.

USE THE TI-83

to draw a histogram to represent the data set on duration of phone calls.

▷▶ *We begin by entering interval midpoints into a "list" called L1. (Each midpoint is a value halfway through the interval; for instance, the first midpoint is 60, halfway between the lower limit of 0 and the upper limit of 120). We will then create a second list to hold the interval frequencies.*

▷▶ *Press 2nd, { , 60, comma, 180, comma, 300, comma, 420, comma, 540, 2nd, } , ENTER, STO, 2nd, L1, ENTER. Then press 2nd, { , 11, comma, 13, comma, 5, comma, 4, comma, 2, 2nd, } , ENTER, STO, 2nd, L2, ENTER.*

▷▶ *Press 2nd, STATPLOT, ENTER, ENTER, down, right, right, ENTER (to pick the histogram icon). Then go down to Xlist and input L1 by pressing 2nd, L1. Then input L2 on the next line by pressing 2nd, L2. Then press WINDOW, and enter these values: 0 for Xmin (lower limit of the first interval), 600 for Xmax (upper limit of the last interval), 120 for Xscl (width of each interval), 0 for Ymin, 13 for Ymax (maximum interval frequency), 5 for Yscl (a reasonable distance between Y-axis tick marks), then press GRAPH to see your histogram.*

▷▶ *To save a graph for later use, press 2nd, Draw, right, right, 1, then pick a number from 0 to 9, ENTER. The histogram is now saved. To recall the histogram at a future time, press 2nd, Draw, right, right, 2, then the proper number, ENTER. As many as 10 graph pictures can be stored in memory.*

2.2 Pie Charts

A *pie chart* is a graphical presentation for showing what percent of the whole each category represents. To construct a pie chart we must have a <u>protractor</u> and a <u>compass</u>. We use the protractor for measuring angles and the compass for drawing a circle. Alternatively, circle templates, made by companies such as Berol Corporation or Pickett Industries, could be used in lieu of a compass to draw a circle.

To illustrate a pie chart, let us reuse the data on bedding sets sold during a holiday weekend sale. As mentioned, pie charts reflect percents of each category rather than the actual count. Therefore, in the table below, we have reproduced the data and the corresponding percents.

Type of Bedding	Number Sold	Percent
Single	8	23.5%
Double	12	35.3
Queen	9	26.5
King	5	14.7
	34	100.0

For example, the percent of single bedding sets sold is (8/34) X 100% = 23.5% rounded to one decimal place.

Each type of bedding will be represented in the pie by a "slice;" the size of the slice depends on the percent of the category. For example, double beds should occupy 35.3% of the pie. One way to apportion 35.3% of the pie to the category is to allocate 35.3% of the 360° in the circle to it. This would translate to a slice with an angle of 35.3% of 360° = .353 times 360° ≈ 127°. (The symbol ≈ means "approximately equal to.")

The angles for each of the other categories would be found in a similar manner:

Single: 23.5% of 360° ≈ 85°
Queen: 26.5% of 360° ≈ 95°
King: 14.7% of 360° ≈ 53°

Next, we use our compass or circle template to draw a circle. Then, starting at the 9 o'clock position (as if the circle were the face of a clock), we draw a horizontal radius line from the edge of the circle to the center of the circle, as indicated below.

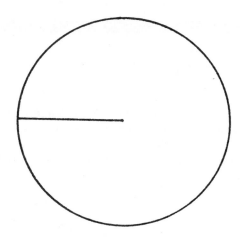

With this line as our base, we use the protractor to measure an angle 127°, corresponding to the largest category. The slice generated by this angle will be assigned to the double bed category.

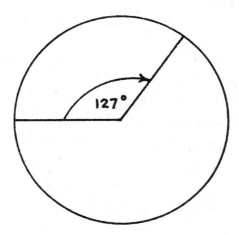

Next, we work with the second largest category of bedding, queen-sized. Using the radius line pointing at about 1 o'clock as the base, we measure an angle of 95°.

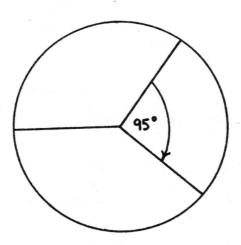

We continue working around the circle, following with the next largest category (single, 85°, and so on) until all the categories have been allocated an angle and a portion of the pie. The final product for these data should look like this (don't forget to label and title the graph):

Bedding Set Sales

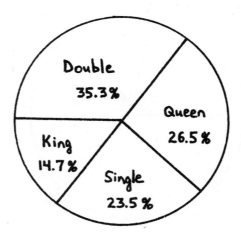

Not all pie charts begin at the 9 o'clock position. Some start at the 12 o'clock position. Also, categories do not have to wrap around the pie in descending order of percent magnitude. However, it is easier to view a pie chart when they do.

 **USING EXCEL**

We can use the **Chart Wizard** key to draw a pie chart for a set of data in an Excel spreadsheet. Let us reuse the bedding sales data set, which you can either retrieve from a file or key-in now. We will assume that the different types of bedding are listed in column A and the number sold are listed in column B.

Click on the **Chart Wizard** key and highlight the word Pie in the Chart type column. In the Chart sub-type box, click on the one called "Pie," which is likely the one in the upper left-hand corner. Click and hold on the button called "Press and hold..." to get a preview of what the pie chart will look like. Then click on Next.

If you have properly entered the data set in columns A and B and rows 2 through 5, then Excel automatically recognizes the source of the data for this graph and displays it in the box called "Data range" in the Step 2 window. If nothing appears in this box, you will need to click and drag the white cross over the cells containing the data. Again, reposition the Chart Wizard window as needed. Click on Next.

In the Step 3 window that appears next, let us change the title of the graph to "Bedding Set Sales." Press the **Tab** key to activate the Chart title box and begin keying-in the new title.

Near the upper right-hand corner of this window are three phrases: Titles, Legend, and Data Labels. Point at the Data Labels phrase and click once. Several options are available for us to add labels to the slices of the pie. Point at the phrase "Show percent" and click once. Notice that percents have appeared near each slice of the pie. Click on Next.

Decide where you wish to paste your newly created pie chart and then click on Finish near the lower right hand corner. Your pie chart should resemble the following figure.

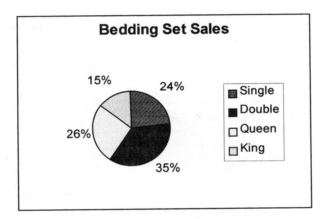

EXCEL F.Y.I.

To create a pie chart...

(1) ... retrieve an existing data set from a file (or key-in a new data set),

(2) click on the **Chart Wizard** key,

(3) select Pie from the Chart types,

(4) select Pie from the Chart sub-types; click Next,

(5) highlight the rows and columns in the spreadsheet where the data reside
   (Note: this may already be done); click Next,

(6) key-in appropriate chart title,

(7) select an appropriate data label option (such as, "Show percents"); click Next,

(8) place chart as an object in your current spreadsheet; click Finish.

SUMMARY EXAMPLE
Pie Chart

Construct a pie chart for these data:

Favorite pizza topping	No. of respondents
Pepperoni (P)	45
Ground beef (GB)	80
Mushrooms (M)	25
Other (O)	20

Do this first:	Find percent for each topping. For example, the percent of pepperoni is (45/170) X 100%= 26.5%	Topping	Number	Percent
		P	45	26.5%
		GB	80	47.0
		M	25	14.7
		O	20	11.8
			170	100%

Do this second:	Find angle for each topping by multiplying percent times 360°. For example for P: 26.5% times 360° ≈ 95°.	Topping	Percent	Angle
		P	26.5%	95°
		GB	47.0	169°
		M	14.7	53°
		O	11.8	43°
			100.0	360°

Do this third:

a. Draw a circle and, starting at the 9 o'clock position, measure the largest angle (169°) with a protractor.

b. Measure the next largest angle (95°), using the radius line pointing at about 3 o'clock as the base.

c. Measure the next largest angle (53°) using the radius line pointing at about 6 o'clock as the base.

d. Label the slices and title the graph.

a.

b.

c.

d.

Favorite Pizza Topping

2.3 Bivariate Graphs

A *bivariate graph* is used to plot paired values of two variables such as X and Y. For instance, suppose the variable X represents the number of weeks of paid vacation per employee and the variable Y represents the number of sick days claimed last year per employee. Further suppose that a sample of 5 employees from a company produced the following data set of paired values for X and Y.

X	Y
1	6
2	3
3	3
4	1
5	1

A plot of these five data points is called a *scatter plot* (because the points are scattered about and may not line up) and is illustrated below.

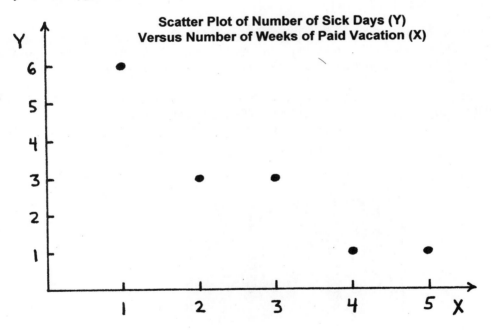

In constructing a scatter plot, we do <u>not</u> connect the points with line segments. The reason we do not is that an objective of an analysis involving two variables such as X and Y is to find the equation of one straight line drawn "through" the scatter plot so as to best describe the relationship of Y to X. This analysis falls under the heading of *regression analysis*.

 **USING EXCEL**

The **Chart Wizard** key allows us to construct a scatter plot for a set of X- and Y-values. Key-in the five pairs of values, found at the beginning of this section, in columns A and B, omitting the labels X and Y. Your data should be in cells A1:B5. Click on the Chart Wizard key and highlight "XY (Scatter)" in the Chart type column.

Select "Scatter" in the Chart sub-type box and click on Next. Excel recognizes the data in columns A and B and displays the active cells in the Data Range box in this Chart Source Data window (Step 2). (If nothing appears in the Data Range box, highlight the appropriate cells in the spreadsheet with the mouse.) Click on Next.

In the Chart Options window (Step 3), let us change the title of the graph to "Scatter Plot of X versus Y." Press the **Tab** key and begin keying in this phrase in the Chart title box. Press the **Tab** key again and key in the letter X in the Category (X) axis box. Then click on Next.

Decide where you wish to paste your newly created scatter plot and then click on Finish near the lower right hand corner. Your graph should resemble the following figure.

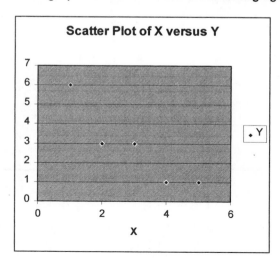

EXCEL F.Y.I.

To create a scatter plot...

(1) ... retrieve an existing data set from a file (or key-in a new data set),
(2) click on the **Chart Wizard** key,
(3) select XY (Scatter) from the Chart types,
(4) select Scatter from the Chart sub-types; click Next,
(5) highlight the rows and columns in the spreadsheet where the data reside
   (Note: this may already be done); click Next,
(6) key-in appropriate chart title and label for X-axis; click Next,
(7) place chart as an object in your current spreadsheet; click Finish.

Continuing with this idea of drawing a straight line, let us review a few properties of lines. The general *equation of a straight line* is:

$$Y = mX + b,$$

where m is the slope and b is the Y-intercept. *Slope*, often defined as "rise-over-run" (or "fall-over-run" if the slope is negative), represents the change in the value of Y for a one-unit change in the value of X. The *Y-intercept* is simply the point of intersection of the vertical axis and the line. Equivalently, the *Y-intercept* is the value of Y when X = 0.

The values of m and b determine the equation of a line. For instance if m = .5 and b = 1, the equation relating Y and X is Y = .5X + 1. To graph this line (or any line for that matter), we need only two points. We recommend using the Y-intercept as one of the points; in this case the point would be (0,1). A second point can be found by picking any value for X, substituting it into the equation and solving for Y. Generally, it is a good idea to select a value of X several units away from the first point, so that when we draw the line connecting the points we have a good spread between them. If we let X = 5, then Y = .5(5) + 1 = 3.5. The graph below shows the two points, the line, the slope, and the Y-intercept.

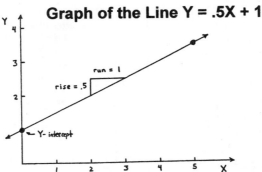

Graph of the Line Y = .5X + 1

Although the equation Y = mX + b is commonly used in algebra texts, most statistics texts use Y = a + bX. There are two points of confusion with regard to the two equations. First, the order of the terms is different: the algebra equation lists the slope component (mX) first and the intercept (b) second. The statistics equation has the intercept (a) first and the slope component (bX) second. The other point of confusion concerns the meaning of the letter b. In the algebra equation, b is the Y-intercept, but in the statistics equation, b is the slope. More than likely your statistics text writes the equation as $Y = a + bX$, so be aware of the difference.

SUMMARY EXAMPLE
Bivariate Graph

Construct a scatter plot for these data and draw the line $Y = -1.8X + 12.6$ on the same graph.		X Y 1 12 2 6 3 9 4 6 5 3
Do this first:	Plot the points and label the axes.	
Do this second:	Choose any value for X, put it into the equation and solve for Y. We chose X = 3.	$Y = -1.8 (3) + 12.6$ $= -5.4 + 12.6 = 7.2$
Do this third:	Plot 12.6, which is the Y-intercept, on the Y-axis. Also plot the point X = 3, Y = 7.2. (Both points are indicated by a box on the graph.) Connect these points with a line, and extend the line past the largest value of X.	Scatter Plot and Graph of the Line $Y = -1.8X + 12.6$

▷▶ USING YOUR TI-83 ▷▶

The TI-83 can do bivariate graphs and draw the straight line that best fits the data -- the regression line -- onto the graph.

‹‹‹

USE THE TI-83
to graph and then fit a line for the data set used in the example immediately above.

▷▶ *Enter the X's into a "list" called L1 by pressing 2ⁿᵈ, {, 1, comma, 2, comma, 3, comma, 4, comma, 5, 2ⁿᵈ, }, ENTER, STO, 2ⁿᵈ, L1, ENTER. Then input the Y's by pressing 2ⁿᵈ, {, 12, comma, 6, comma, 9, comma, 6, comma, 3, 2ⁿᵈ, }, ENTER, STO, 2ⁿᵈ, L2, ENTER.*

▷▶ *Now press 2ⁿᵈ, STAT PLOT, ENTER, ENTER, WINDOW. On the WINDOW menu, we need to pick some options (use the down arrow key to move around): choose 0 for Xmin (a number just less than the smallest X), choose 6 for Xmax (a number just more than the largest X), let Xscl (this defines the difference between tick marks on the X-axis) be 1, Ymin = 0, Ymax = 13, then press GRAPH.*

▷▶ *To plot the best fit line onto this graph, press STAT, right arrow, 4, 2ⁿᵈ, L1, comma, 2ⁿᵈ, L2, comma, VARS, right arrow, 1, 1, ENTER, GRAPH. If you also want to label the axes, press 2ⁿᵈ, FORMAT, then use the arrow keys to highlight "LabelOn," then press ENTER, GRAPH. Press CLEAR if you wish to see the line equation.*

‹‹‹

Another use of bivariate graphs will occur when we construct a *time series graph*. Again we will be dealing with two variables, labeled t and Y_t in this application. The variable t represents time and is usually scaled in integers, starting at 1 and counting forward. Consider, for example, the following data set where Y_t = the number of mortgage loans approved per month by a small bank:

Month	t	Y_t
January	1	12
February	2	10
March	3	19
April	4	24
May	5	32
June	6	28

Month	t	Y_t
July	7	33
August	8	21
September	9	15
October	10	20
November	11	18
December	12	8

To graph these time series data we would plot t on the horizontal axis and Y_t on the vertical axis as indicated below. Unlike the scatter plot where we did not connect the points, the time series graph shows line segments connecting each successive observation. In this way the graph takes the appearance of a sequence of jagged lines. One of our objectives in time series analysis will be to try to discover any trends or patterns within such a sequence.

Number of Mortgage Loans Approved per Month

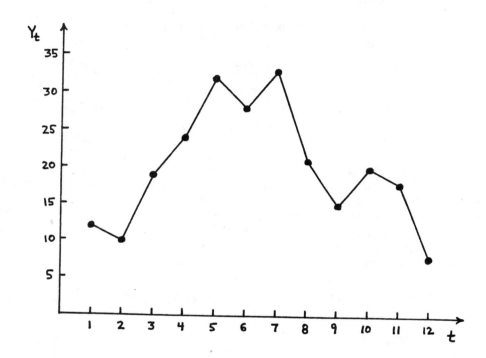

Data analysis always should begin with a well-constructed graph. You may discover that your graph may get more scrutiny than the written analysis that accompanies it! Make the graph a professional masterpiece.

▷▶ USING YOUR TI-83 ▷▶

The TI-83 can display a *time series graph*. In the example below we'll also see how to give a new data set (a "list," in TI-83 terminology) a descriptive name. Also we'll see how to "paste" a list name to a screen. (The LIST key is above the STAT key.)

〰〰〰〰〰〰〰〰〰〰〰〰〰〰〰〰〰〰〰〰〰〰

USE THE TI-83 to generate a time series graph for the loans approved series given above.

▷▶ *To input the time periods, press 2nd, { , 1, comma, 2, comma, •••, 11, comma, 12, 2nd, } , STO, ALPHA, T, ALPHA, I, ALPHA, M, ALPHA, E, ENTER. To enter the loans, press 2nd, { , 12, comma, 10, comma, •••, 18, comma, 8, 2nd, } , STO, ALPHA, L, ALPHA, O, ALPHA, A, ALPHA, N, ALPHA, S, ENTER. (If you wish to view a listing of named data sets in the TI-83 memory, press 2nd, LIST.)*

▷▶ *Now press 2nd, STAT PLOT, ENTER, ENTER, down, right, ENTER, then down to the Xlist line. To avoid using the ALPHA key to re-specify our two lists TIME and LOANS, press 2nd, LIST, 2, down, 2nd, LIST, 1. (The previous sentence assumes TIME is list 2 and LOANS is list 1. If they have different numbers, adjust accordingly.) Then press WINDOW and choose Xmin = 0, Xmax = 13, Xscl = 1, Ymin = 0, Ymax = 35, Yscl = 1, and then touch GRAPH to see the final product.*

〰〰〰〰〰〰〰〰〰〰〰〰〰〰〰〰〰〰〰〰〰〰

Chapter 2 Exercises

2.1 There are three forms of business ownership: sole proprietorship, partnership, and corporation. A survey of 50 small businesses revealed the following breakdown.

Form of Business	Number
Sole proprietorship	17
Partnership	22
Corporation	11

Draw a bar graph to represent these data.

2.2 Refer to Exercise 2.1. Construct a pie chart for the data.

continued on the next page...

Chapter 2 Exercises (continued)

2.3 Graph the following probability distribution.

x	P(X = x)
1	.40
2	.25
3	.15
4	.20

2.4 Reconstruct the scatter plot of Y = number of sick days claimed last year per employee versus X = number of week's paid vacation per employee, and then draw the line Y = –1.2X + 6.4 on the scatter plot. The data appear at the beginning of Section 2.3. Does the line touch any of the points?

2.5 The most recent quarterly dividends paid on a share of common stock by each company in a sample of 23 retailing businesses produced the following distribution. Construct a histogram for these data.

Dividends	Number
$0.00 -- 0.85	4
0.85 -- 1.70	9
1.70 -- 2.55	7
2.55 -- 3.40	3

2.6 The quarterly earnings-per-share of a computer company's stock are listed below for a two-year period.

Year	Quarter	Earnings-per-Share	Year	Quarter	Earnings-per-Share
1	1	$.46	2	1	$.46
	2	.25		2	.26
	3	.24		3	.40
	4	.25		4	.53

Construct a bivariate graph for these time series data.

2.7 The broad categories of graphs in business statistics are bar graphs, pie charts, and bivariate graphs. Into which category do each of the following graphs fall?

(a) scatter plot
(b) histogram
© time series graph
(d) discrete probability distribution

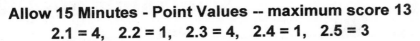

POSTTEST #2

Allow 15 Minutes - Point Values -- maximum score 13
2.1 = 4,   2.2 = 1,   2.3 = 4,   2.4 = 1,   2.5 = 3

2.1 Construct a histogram for these data.

Number of hours spent watching TV per day	Number of people responding
0 - under 3	5
3 - under 6	25
6 - under 9	20
9 - under 12	10

2.2 Which type of graph -- bar graph, pie chart, or bivariate graph -- is used to draw time series data?

2.3 Contrast a bar graph versus a discrete probability distribution in terms of:

   (a) the numbers plotted on the vertical axis.

   (b) the gaps between the rectangles along the horizontal axis.

2.4 True or false: In drawing a scatter plot we should connect the points with a series of line segments.

2.5 Consider the line Y = 1.6X + 4.

   (a) What is the value of the slope?

   (b) What is the value of the Y-intercept?

   (c) Does the graph of the line go through the point X = 5, Y = 12?

PRETEST #3

Allow 20 Minutes
1 point each -- maximum score 10

3.1 What symbol is used as a shorthand notation to represent the process of summing numbers?

**QUESTIONS 3.2 - 3.6 REFER TO THE FOLLOWING
SET OF DATA. COMPUTE THE INDICATED QUANTITIES.**

$$\underline{X}$$
1
7
0
8
−6
―――

3.2 ΣX

3.3 ΣX^2

3.4 $(\Sigma X)^2$

3.5 $\Sigma(X - 2)$

3.6 Sum of the squared deviations (about the mean)

3.7 Translate the following phrase into summation notation: the sum of the squared X-values.

**QUESTIONS 3.8 - 3.10 REFER TO THE FOLLOWING
SET OF DATA. COMPUTE THE INDICATED QUANTITIES.**

X	Y
12	2
10	-1
5	5

3.8 ΣXY

3.9 $\overline{Y}$

3.10 $\Sigma(X - \overline{X})(Y - \overline{Y})$

Years ago we encountered road signs along the highway with printed messages such as "No Left Turn" or "Slippery When Wet." Words have given way to pictures that depict the prohibition of turning left, such as the following one.

The circle with the line through it has become a universally accepted and understood symbol.

Similarly, we have symbols in statistics that are used to represent certain mathematical operations. A common one is the upper case Greek letter *sigma* (Σ). It is used to symbolize addition (or summation). Certain rules and conventions have evolved with the sigma symbol. First is the literal translation of the symbol Σ in English as "the summation of." The summation of <u>what</u>, you might ask? The answer is the numbers that result from sampling, which represent the values of a variable.

For example, suppose we sampled six members of our class and asked each to indicate how many telephones they have at home. Further suppose the responses were as follows: 2, 1, 2, 3, 3, and 1. Just as a retail business has an accounting system to record revenues and expenses, statistics has developed an "accounting system" to track data that result from sampling. We begin by defining, in words, the variable under study; in this case, we let X = the number of telephones per residence. Once X has been defined we can write our sample data set in a column with X as the heading:

$$\frac{X}{\begin{array}{c} 2 \\ 1 \\ 2 \\ 3 \\ 3 \\ 1 \end{array}}$$

3.1 Sum of the Values

By writing the values of X in a column it is easy to perform the arithmetical operation of addition. For instance, when we add the six numbers in the X column below we get 12, which is called the *sum of the values* and associated with the symbol ΣX.

$$\frac{X}{\begin{array}{c} 2 \\ 1 \\ 2 \\ 3 \\ 3 \\ \underline{1} \\ 12 \end{array}} = \Sigma X$$

The sum of the values is used to find the *mean* of a data set, which represents the average value in the data set. For a sample data set we use the symbol $\overline{X}$ to denote the mean and the symbol n to indicate the number of values in the data set. The formula for the mean follows.

«FORMULA»
MEAN

$$\overline{X} = \frac{\Sigma X}{n}$$

In the previous example where X was defined as the number of telephones per residence, we had $\Sigma X = 12$ and $n = 6$. Thus, the mean is 2 telephones per residence:

$$\overline{X} = \frac{\Sigma X}{n} = \frac{12}{6} = 2$$

	SUMMARY EXAMPLE Sum of the Values	
For these data, compute (1) ΣX and (2) $\overline{X}$:		$\begin{array}{c} \underline{X} \\ 5 \\ -3 \\ 1 \\ \underline{9} \end{array}$
Part (1):	Add the numbers	$\begin{array}{c} \underline{X} \\ 5 \\ -3 \\ 1 \\ \underline{9} \\ 12 = \Sigma X \end{array}$ **Answer↗**
Part (2):	Divide the total by n	$\overline{X} = \frac{\Sigma X}{n} = \frac{12}{4} = 3$ **← Answer**

Summation Notation

In general, the column heading becomes the *variable of summation* when the values in that column are to be added. The total is recorded at the bottom of the column and denoted by combining the symbol Σ with the variable of summation. For instance, if the values under the column heading X – 2 are added, the variable of summation is X – 2 and the total at the bottom of the column is denoted by Σ(X – 2). Notice that we put parentheses around the variable of summation X – 2. This is because there is a difference between the symbols Σ(X – 2) and ΣX – 2, as we will explain shortly.

 ▷▶ USING YOUR TI-83 ▷▶

The TI-83 can quickly determine ΣX and X̄ for a given data set. We show how for the four-item data set given above.

∿∿∿∿∿∿∿∿∿∿∿∿∿∿∿∿∿∿∿∿∿∿∿∿∿∿∿∿∿∿∿∿∿

USE THE TI-83   **Enter the values into the TI-83 memory by pressing 2ⁿᵈ, { , 5, comma, (-) 3, comma, 1, comma, 9, 2ⁿᵈ, } , STO, ALPHA, (letter) X, ENTER. We now have entered a four-item data set named X into memory.**

▷▶ *To get ΣX, press 2ⁿᵈ, LIST, left arrow, 5, 2ⁿᵈ, LIST, 1 (or whatever number data set X is listed as), ENTER.*

D I S P L A Y: **12, which is ΣX**

▷▶ *To get the mean, X̄, press 2ⁿᵈ, LIST, left arrow, 3, 2ⁿᵈ, LIST, 1 (or whatever), ENTER.*

D I S P L A Y: **3, which is the mean, X̄**

An alternative way to get X̄ and ΣX, given that the data set has been entered into memory, would be to press STAT, right arrow, 1, 2ⁿᵈ, LIST, 1 (or whatever), ENTER. This generates a lot of statistics -- the first is X̄; the second is ΣX.

∿∿∿∿∿∿∿∿∿∿∿∿∿∿∿∿∿∿∿∿∿∿∿∿∿∿∿∿∿∿∿∿∿

 **USING EXCEL**

Sum of the Values

Excel is capable of determining the sum of the values with one keystroke. First, let's create a data set, just like the one above in the Summary Example. Key in the letter "X" in cell A1 and then enter the values 5, -3, 1, and 9 in cells A2 through A5, respectively. Save the data set – call the file "Chap 3 data," for example.

Now make cell A6 the active cell and then scan the standard toolbar (the third line from the top of the screen) for the Greek letter sigma, Σ, which represents the **AutoSum** key. Click on this symbol and notice that the notation "=SUM(A2:A5)" appears in the status line. Excel automatically recognizes the cells in which you have entered data and is ready to add up the values for you. Press the Enter key or click on the green check mark in the status line. The sum of "12" should now appear in cell A6.

EXCEL F.Y.I.

To automatically add the values in a column, activate the empty cell immediately beneath the column of numbers, click on the Σ key, and then press the Enter key.

Mean

To calculate the mean of a set of numbers we can use the **Paste Function** feature in Excel. (Note: In previous versions of Excel this feature was called "Function Wizard.")

For example, let us find the mean of the data set in the Summary Example. Scan the standard toolbar again for the symbol f_x. It should be immediately to the right of the Σ symbol. Make cell A7 active and then click on the f_x symbol. Your screen should resemble the following figure.

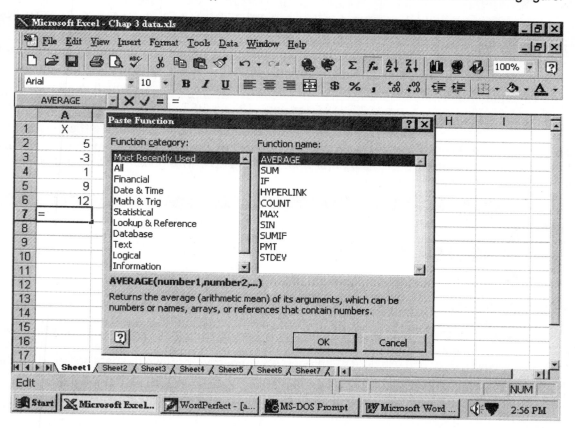

Highlight the word All in the Function Category list. Then scroll down the Function Name list until you find the word AVERAGE. (In this application, the context of the words "average" and "mean" is identical, so determining the <u>mean</u> of a set of numbers is the same as determining the <u>average</u>.)

If not already highlighted, highlight the word AVERAGE in the second column. Click on the word "OK" near the lower right-hand corner of this screen. (If you are using an earlier version of Excel, click on the word "Next.")

We need to tell Excel which numbers are to be averaged. (If the Paste Function window, shown in the preceding figure, is covering up your data in column A, you will need to reposition it so that your data are visible. Click, hold, and drag the Paste Function window to a new position.)

Now move the mouse so that the white cross is positioned over cell A2 and then click, hold, and highlight cells A2 through A5. Release the mouse button. The notation "=AVERAGE(A2:A5)" should appear in the status line and your screen should resemble the following figure.

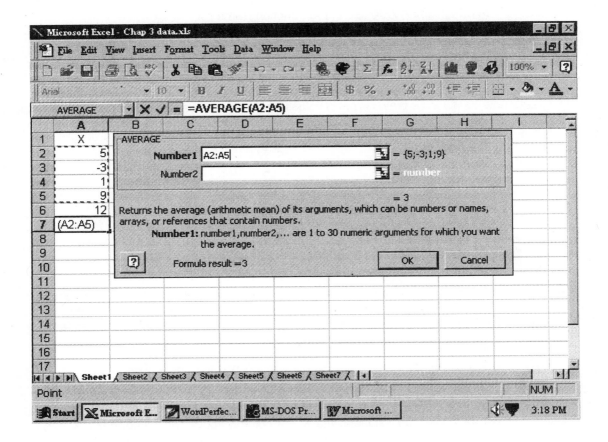

Click on the word "OK" in the lower right corner of the Paste Function window when your screen imitates the preceding figure. (If you are using an earlier version of Excel, click on the word "Finish.")

The number 3 should appear in cell A7 and this number is the average or mean for our data set.

EXCEL F.Y.I.

To find the mean of a set of numbers...

    (1)   ...make an empty cell active,

    (2)   click on the Paste Function key (f_X),

    (3)   highlight All in the Function Category list,

    (4)   highlight AVERAGE in the Function Name list,

    (5)   click on the word OK in the Paste Function window,

    (6)   highlight the cells of the data to be averaged, and

    (7)   click on the word OK.

Let's now consider the previous set of values for X = the number of telephones per residence.

X	(X – 2)
2	0
1	-1
2	0
3	1
3	1
1	-1
12 = ΣX	0 = Σ(X – 2)

We generated the second column with the heading X – 2 by subtracting 2 from each X-value in the first column. Notice that the sum of the (X – 2) values is 0 and that the summation notation for this total is Σ(X – 2) = 0. Conversely, if we write ΣX – 2, then, in words, we are telling the reader to add the values in the column headed by X, and then subtract 2 from the total: ΣX – 2 = 12 – 2 = 10.

Because we found the mean of the previous data set to be $\overline{X}$ = 2, the column labeled X – 2 above also could have been labeled as X – $\overline{X}$. The expression X – $\overline{X}$ is given a special name in statistics: *deviation*. (The formal term is *deviation from the mean*, but we usually refer to it as (just) *deviation*.) When we added the values in the X – 2 column above, we found that Σ(X – 2) = 0. Equivalently, this sum could be written as Σ(X – $\overline{X}$) = 0. The symbol Σ(X – $\overline{X}$) is called the *sum of the deviations*.

WORD PROBLEM EXAMPLE

In Joshua's statistics class there were five scheduled quizzes during the term. The maximum possible score on each quiz was 20 points. He made perfect scores on the first and the last quiz, missed the second one, and scored 17 and 15 on the third and fourth quiz. What is his average quiz score?

Problem Solving Diagnostics (see page 14):

1.	**Key words:**	five scheduled quizzes; maximum possible score; perfect scores; average quiz score
2.	**Numbers:**	5 = number of quizzes 20 = score on 1st quiz 0 = score on 2nd quiz 17 = score on 3rd quiz 15 = score on 4th quiz 20 = score on 5th quiz
3.	**Picture:**	Not necessary
4.	**Question mark:**	What is the value of $\overline{X}$?
5.	**Computation:**	

$$\overline{X} = \frac{\Sigma X}{n} = \frac{20 + 0 + 17 + 15 + 20}{5} = \frac{72}{5} = 14.4$$

Joshua's average quiz score was 14.4

3.2 Sum of the Squared Values

As the previous examples demonstrate, the variable of summation can be X or X – 2, or even X^2. If we were to write the symbol ΣX^2, then the term is called the *sum of the squared values*. The variable of summation is X^2, and as we learned in Chapter 1, the operation of exponentiation (squaring) takes precedence over the summation. Thus, we should square the X-values first, and then add the squared values. We believe it is easier to see this process unfold if we put the squared numbers in a separate column labeled X^2. Let us use the previous sample data for X and find ΣX^2.

X	X^2
2	4
1	1
2	4
3	9
3	9
1	1
12 = ΣX	28 = ΣX²

The sum of the X^2-values is 28. Again note that the symbol to denote 28 -- ΣX^2 -- follows our convention of combining the symbol Σ with the column heading X^2.

SUMMARY EXAMPLE
Sum of the
Squared Values

Compute ΣX^2 for these data:

X
5
-3
1
9

Do this first:	Square each number	$5^2 = 25$, $(-3)^2 = 9$, $1^2 = 1$, $9^2 = 81$
Do this second:	Add the squares	X^2 25 9 1 81 $\Sigma X^2 = 116$ ← Answer

---

Similarly, we might be interested in the sum of the squared (X − 2) values. Recreating the (X − 2) values and squaring each value in a separate column, we have:

X	(X – 2)	(X – 2)²
2	0	0
1	-1	1
2	0	0
3	1	1
3	1	1
1	-1	1
	$0 = \Sigma(X - 2)$	$4 = \Sigma(X - 2)^2$

The notation at the bottom of the $(X - 2)^2$ column is consistent with our general guideline that combines the Σ symbol with the column heading. Again, because $\overline{X} = 2$ we could have labeled the columns in the above example $X - \overline{X}$ and $(X - \overline{X})^2$ and written the totals of the columns as $\Sigma(X - \overline{X}) = 0$ and $\Sigma(X - \overline{X})^2 = 4$. The symbol $\Sigma(X - \overline{X})^2$ is called the *sum of the squared deviations*.

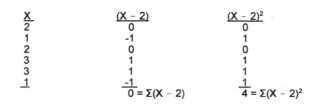

▷▶USING YOUR TI-83 ▷▶

To get ΣX^2 we will use the "1-variable statistics" function on the TI-83. We'll use the small data set above ($X = 5, -3, 1,$ and 9) to illustrate.

USE THE TI-83   **Enter the data set and name it: press 2ⁿᵈ, { , 5, comma, (-) 3, comma, 1, comma, 9, 2ⁿᵈ, } , STO, ALPHA, X, ENTER.**

▷▶ *Now press STAT, right arrow, 1, 2ⁿᵈ, LIST , 1 (or whatever number data set X is listed as), ENTER. The third quantity displayed is $\Sigma X^2 = 116$.*

USING EXCEL

The Paste Function key may also be used to calculate the sum of the squared values. We will illustrate with the data from the above Summary Example. (If you saved this data in a file called "Chap 3 data," retrieve it now.)

Make cell A8 the active cell and click on the f_x key. In the left-hand column labeled Function Category, highlight All.

An alphabetized list of functions appears in the right-hand column labeled Function Name. Scroll down this list until you see the function called SUMSQ and then highlight this term. Then click on the word OK (or the word NEXT, if you are using an earlier version of Excel) near the lower right-hand corner of the Paste Function window.

Move the mouse so that the white cross is posititioned over cell A2 and then click and drag over cells A2 through A5. Release the mouse button. The notation "=SUMSQ(A2:A5)" should appear in the status line. Click on the word OK (or the word FINISH, if you are using an earlier version of Excel) in the lower right hand corner of the Paste Function window. The resulting calculation of 116 should now appear in cell A8.

EXCEL F.Y.I.

To find the sum of the squared values, ΣX^2, for a set of numbers ...

   (1)   ... make an empty cell active,
   (2)   click on the Paste Function key (f_x),
   (3)   highlight All in the Function Category list,
   (4)   highlight SUMSQ in the Function Name list,
   (5)   click on the word OK in the Paste Function window,
   (6)   highlight the cells of the data to be squared, and
   (7)   click on the word OK.

3.3 Sum of the Values Squared

Now suppose that we were to write $(\Sigma X)^2$, which is translated as the *sum of the values squared*, or as the *square of the sum of the values*. Again, from Chapter 1, we know that the parentheses direct us to perform the operation within the parentheses <u>before</u> performing any other operation. Therefore, we should find ΣX first, and then square. Earlier we computed $\Sigma X = 12$; thus, $(\Sigma X)^2 = 12^2 = 144$. Notice that $(\Sigma X)^2$ is <u>not</u> the same as ΣX^2.

		SUMMARY EXAMPLE **Sum of the Values Squared**
Compute $(\Sigma X)^2$ for these data:		X 5 -3 1 9
Do this first:	Add the numbers	X 5 -3 1 9 $\overline{12 = \Sigma X}$
Do this second:	Square the total	$(\Sigma X)^2 = (12)^2 = $ **144 ← Answer**

USE THE TI-83

We have shown in the two previous TI-83 examples two different ways to establish that $\Sigma X = 12$. Since we need to square this value, do this:

▷▶ *Input 12, press the x^2 key, and then the ENTER key.*

🄳🄸🄎🄿🄻🄰🅈🄴 144, which is $(\Sigma X)^2 = (12)^2$

3.4 Sum of the Products

One other summation that we will encounter in the early going involves the product of two variables such as X and Y. Consider the following data set:

X	Y
2	2
1	2
2	1
3	2
3	3
1	1

The symbol ΣXY, which is called the *sum of the products*, suggests that the variable of summation is XY, whose values are obtained by multiplying each X-value and its corresponding Y-value. The values are listed below in a separate column labeled XY.

X	Y	XY
2	2	4
1	2	2
2	1	2
3	2	6
3	3	9
1	1	1
		24

The total of the XY column is 24, denoted as $\Sigma XY = 24$.

Summation Notation

SUMMARY EXAMPLE
Sum of the Products

Compute ΣXY for these data:	X	Y
	5	2
	-3	-2
	1	4
	9	0

Do this first:	Multiply each pair	5 · 2 = 10; (-3) (-2) = 6; 1 · 4 = 4; 9 · 0 = 0
Do this second:	Add the products	XY 10 6 4 0 ΣXY = **20** ← **Answer**

 ▷►**USING YOUR TI-83** ▷►

Once we enter the data into the TI-83, it is a simple matter to get ΣXY.

~~~~~~

USE THE TI-83   **Enter the X column by pressing 2ⁿᵈ, { , 5, comma, (-) 3, comma, 1, comma, 9, 2ⁿᵈ, } , STO, ALPHA, X, ENTER. Enter the Y column by pressing 2ⁿᵈ, { , 2, comma, (-) 2, comma, 4, comma, 0, 2ⁿᵈ, } , STO, ALPHA, Y, ENTER.**

▷► *To get ΣXY, press STAT, right arrow, 2, 2ⁿᵈ, LIST, 1 (or whatever number goes with X), comma, 2ⁿᵈ, LIST, 2 (or whatever number goes with Y), ENTER. The resulting display gives selected statistics; you'll need to use the down arrow key a few times until you find ΣXY = 20.*

~~~~~~

Summation Notation

 **USING EXCEL**

To use Excel to compute the sum of the products we must create a second array of data – the Y-values. Open the data file called "Chap 3 data" that you saved to disk on drive A. This file should contain the X-values in cells A2 through A5. Make cell B1 the active cell and enter the letter "Y" in this cell. Then key in the values 2, -2, 4, and 0 in cells B2 through B5, respectively.

Make cell B7 the active cell and click on the Paste Function key, f_x. Highlight All in the Function Category list on the left and then scroll down the list of functions on the right side until you find the one called SUMPRODUCT. Highlight this term in the Function Name list and then click on the word OK near the lower right-hand corner.

You may need to reposition the SUMPRODUCT window so that all the numbers in columns A and B are visible. The cursor should be blinking in a box called "Array 1" in the SUMPRODUCT window. Move the mouse so that the white cross is positioned over cell A2. Then highlight cells A2 through A5 by clicking, holding, and dragging the white cross over those cells. Release the mouse button. The notation "A2:A5" should appear in the white box associated with the term "Array 1."

Press the **Tab** key to activate the "Array 2" box. Move the white cross over the spreadsheet and highlight cells B2 through B5. Release the mouse button. Then click on the word OK in the SUMPRODUCT window. The number 20 should appear in cell B7.

EXCEL F.Y.I.

To find the sum of the products, ΣXY, for a set of X-values and Y-values ...

(1) ... make an empty cell active,

(2) click on the Paste Function key (f_x),

(3) highlight All in the Function Category list,

(4) highlight SUMPRODUCT in the Function Name list,

(5) click on the word OK in the SUMPRODUCT window,

(6) highlight the cells containing the X-values for Array 1,

(7) highlight the cells containing the Y-values for Array 2, and

(8) click on the word OK.

3.5 Sum of the Squares for X

Just as the Greek letter sigma (Σ) is a convenient shorthand symbol to summarize the operation of adding numbers together, we often attach symbols to other terms and formulas throughout statistics. For instance, there is a particular combination of operations that involve ΣX and ΣX^2 to yield a key quantity in descriptive statistics and in regression called the *sum of the squares for X*. We use the letters SSX to symbolize the sum of the squares for X.

《*FORMULA*》
SUM OF SQUARES FOR X

$$SSX = \Sigma X^2 - \frac{(\Sigma X)^2}{n}$$

or

$$SSX = \Sigma (X - \overline{X})^2$$

You will see two formulas in the previous box. Both give the same answer, so you can pick which one you find easier to use. You might recognize the latter one; we earlier referred to it as the sum of the squared deviations. Take note! These terms are the same: *sum of the squared deviations* and *sum of the squares for X*.

The symbol SSX is not common to every statistics text. Your book might use SS(X) or SS_x.

SUMMARY EXAMPLE
Sum of the Squares for X

Compute SSX for these data:		$$\frac{X}{\begin{array}{c}5\\-2\\0\\9\end{array}}$$
Do this first:	Add the numbers	$\frac{X}{\begin{array}{c}5\\-2\\0\\9\end{array}}$ $12 = \Sigma X$
Do this second:	Square each number and then add	$\frac{X^2}{\begin{array}{c}25\\4\\0\\81\end{array}}$ $110 = \Sigma X^2$
Do this third:	Square the first total and divide by n	$\frac{(\Sigma X)^2}{n} = \frac{(12)^2}{4} = 36$
Do this fourth:	Subtract	$SSX = \Sigma X^2 - \frac{(\Sigma X)^2}{n}$ $= 110 - 36 = \mathbf{74}$ **Answer ↑**

Notice in the example that one of the numbers in the X column was 0. We <u>cannot</u> delete this value just because it's 0. We count it as one of the n = 4 points.

USING YOUR TI-83 ▷▶

The TI-83 is not programmed to compute and report SSX, as it does for other sums like ΣX or ΣX^2. But we can still get SSX; we will need to specify and solve an equation to do so.

USE THE TI-83

To replicate the example above, begin by inputting the values: **press 2nd, { , 5, comma, (–)2, comma, 0, comma, 9, 2nd, } , STO, ALPHA, Z (an arbitrary choice), ENTER.**

▷▶ *Press STAT, right arrow, 1, 2nd, LIST, 1 (or whatever number data set Z is listed as), ENTER. We now have the two quantities we need to continue: $\Sigma X = 12$, and $\Sigma X^2 = 110$.*

▷▶ *Now input 12, STO, ALPHA, S, ENTER, to define $S = 12 = \Sigma X$.*

▷▶ *Now input 110, STO, ALPHA, T, ENTER, to define $T = 110 = \Sigma X^2$. (The choice of S and T are arbitrary here.) Now input 4, STO, ALPHA, N, ENTER to define $N = 4$ = sample size.*

▷▶ *Now we can do the arithmetic in the bottom box of the previous example: press ALPHA, T, minus, ALPHA, S, X^2, division key, ALPHA, N, ENTER.*

DISPLAY: **74, which is SSX**

 **USING EXCEL**

Excel has a built-in function for computing SSX, the sum of the squares for X. To demonstrate how to use this function, let us key in the data in the Summary Example (5, -2, 0, and 9) in cells A2 through A5, respectively, and the letter X in cell A1. Make cell A7 the active cell and click on the Paste Function key, f_x.

Summation Notation

Highlight All in the Function Category list on the left side of the Paste Function window and then scan the Function Name list on the right side for the term DEVSQ. Highlight this term and click on OK.

Click and highlight the four numbers residing in cells A2 through A5. Then click on the word OK in the lower right-hand corner of the DEVSQ window. The number 74 should appear in cell A7.

EXCEL F.Y.I.

To find the sum of the squares for X, SSX, for a set of numbers…

(1)   … make an empty cell active,
(2)   click on the Paste Function key (f_x),
(3)   highlight All in the Function Category list,
(4)   highlight DEVSQ in the Function Name list,
(5)   click on the word OK in the Paste Function window,
(6)   highlight the cells containing the data, and
(7)   click on the word OK.

3.6   Sum of the Crossproducts for X and Y

If we have two variables X and Y, we define the *sum of the crossproducts for X and Y* according to the following formulas. The commonly used symbol is either SSXY (or SXY).

〜〜〜〜〜〜〜〜〜〜〜〜〜〜〜〜〜〜〜〜〜〜〜〜
〜〜〜〜〜〜〜〜〜〜〜〜〜〜〜〜〜〜〜〜〜〜〜〜

«FORMULA»
SUM OF THE CROSSPRODUCTS

$$SSXY = \Sigma XY - \frac{(\Sigma X)(\Sigma Y)}{n}$$

$$\textbf{or} \quad SSXY = \Sigma(X - \bar{X})(Y - \bar{Y})$$

〜〜〜〜〜〜〜〜〜〜〜〜〜〜〜〜〜〜〜〜〜〜〜〜
〜〜〜〜〜〜〜〜〜〜〜〜〜〜〜〜〜〜〜〜〜〜〜〜

Summation Notation

To use the first formula to compute SSXY, we recommend that you use three columns--one labeled X, one labeled Y, and one labeled XY. Find the total of each column and use the totals as indicated in the first formula. In the second formula the symbol $\overline{Y}$ is the mean of the Y-values:

$$\overline{Y} = \frac{\Sigma Y}{n}$$

You likely will use summation notation extensively throughout your statistics text. Our discussion here is necessarily brief, but sufficient for the uses of Σ in most texts. Please be aware, however, that the symbol is a condensed version of more explicit notations involving an *index*, denoted by *i*, such as ΣX_i, or even

$$\sum_{i=1}^{6} X_i$$

We have not discussed the latter notation since it is more appropriate for "partial sums," those involving some, but not all, of the sample values. Since most summations include all values, we have chosen to write simply ΣX, with the understanding that no values are to be omitted.

EXCEL FUNCTION CONVERSION CHART

ENGLISH PHRASE	SYMBOL	EXCEL FUNCTION
Sum of the values	ΣX	SUM (or Σ key)
Mean	$\overline{X}$	AVERAGE
Sum of the squared values	ΣX^2	SUMSQ
Sum of the products	ΣXY	SUMPRODUCT
Sum of the squares for X	SSX	DEVSQ

Chapter 3 Exercises

EXERCISES 3.1 - 3.6 REFER TO THE FOLLOWING SET OF DATA.
COMPUTE THE INDICATED QUANTITIES.

$$\begin{array}{r} \underline{X} \\ 5 \\ 4 \\ -10 \\ 1 \\ 6 \\ -4 \\ 4 \\ 8 \\ 0 \\ \underline{6} \end{array}$$

3.1 ΣX

3.2 $\overline{X}$

3.3 ΣX^2

3.4 $(\Sigma X)^2$

3.5 $\Sigma(X - 7)$

3.6 $\Sigma(X - 2)^2$

3.7 Express the following phrases in summation notation:

(a) the sum of the squared values of the variable X.
(b) the square of the sum of the values of the variable X.
(c) the sum of the values of the variable X, minus 10.
(d) the sum of the values of the variable X − 10.
(e) the sum of the product of the corresponding values of the variables X − 3 and Y − 2.

EXERCISES 3.8 - 3.13 REFER TO THE FOLLOWING SET OF DATA.
COMPUTE THE INDICATED QUANTITIES.

$$\begin{array}{r} \underline{X} \\ 8 \\ 4 \\ 6 \\ 4 \\ \underline{8} \end{array}$$

3.8 $\Sigma X - 50$

3.9 $\Sigma(X + 5)$

3.10 $(\Sigma X)^2$

3.11 ΣX^2

3.12 sum of the deviations

3.13 sum of the squared deviations

continued on the next page ...

Chapter 3 Exercises (continued)

3.14 Name the term that is associated with each symbol:

(a) $\overline{X}$

(b) $X - \overline{X}$

(c) ΣXY

(d) SSX

(e) ΣX^2

EXERCISES 3.15 - 3.21 REFER TO THE FOLLOWING SET OF DATA.
COMPUTE THE INDICATED QUANTITIES.

X	Y
4	−1
0	−1
2	−1
2	4
−3	1
1	1

3.15	$\overline{X}$	**3.19** $\Sigma(X - 2)(Y - 1)$	
3.16	$\overline{Y}$	**3.20** SSX	
3.17	ΣXY	**3.21** SSXY	
3.18	$(\Sigma X)(\Sigma Y)$		

3.22 Suppose you had a table of numbers with the following column headings:

$$\underline{f \quad M \quad fM \quad fM^2}$$

If the total of the numbers in the respective columns (from left to right) were 20, 40, 179, and 1733, what symbol would you use to denote the number...

(a) 40? (b) 1733? (c) $(179)^2$?

3.23 A telephone survey of local computer stores was conducted to compare the price of a new video game. Two stores reported a price of $49. Another store charged $59. The last store called had a special sale price on the video game of $43.

(a) What is the average selling price?

(b) Find the value of the sum of the squares for X, the price of the video game.

Summation Notation

POSTTEST #3

Allow 20 Minutes
Point values -- maximum score 10
3.1 = 6, 3.2 through 3.5 = 1 point each

3.1 Find the term on the right that matches the symbol on the left:

Symbols		Terms
(a)	$\Sigma(X - \overline{X})$	1. mean
(b)	ΣX^2	2. summation
(c)	$\overline{X}$	3. sum of the values
(d)	SSX	4. index
(e)	$(\Sigma X)^2$	5. covariance
(f)	ΣXY	6. sum of the values squared
		7. deviation
		8. sum of the products
		9. sum of the squared values
		10. sum of the deviations
		11. sum of the squares for X
		12. sum of the crossproducts for X and Y

QUESTIONS 3.2 - 3.5 REFER TO THE FOLLOWING SET OF DATA.
COMPUTE THE INDICATED QUANTITIES.

X	Y
5	3
9	1
-4	6
6	2

3.2 $\overline{X}$

3.3 $\Sigma(X - \overline{X})^2$

3.4 ΣXY

3.5 SSXY

PRETEST #4
Allow 15 Minutes
Calculator with an x^y key needed
1 point each part -- maximum score 10

4.1 Write the following as x to a single exponent:

    (a) $x(x^4)(x^2)$

    (b) $(x^8)^5$

    © $\dfrac{x^4}{x^9}$

4.2 Write the following using the radical symbol ($\sqrt{}$):

    (a) $x^{1/2}$ **(b)** $x^{1/3}$ © $x^{2/5}$

4.3 Use your calculator to find the following numbers; round off to 4 decimal places, as needed:

    (a) 4^7 **(b)** 5^{-4} © $\sqrt[5]{6}$

4.4 A calculator computation yielded the display 3.2 $^{-04}$. What number is being represented by this display?

4.5 Write the following numbers in scientific notation:

    (a)   1,953,000

    (b)   .000577

        The mathematical operations of squaring a number and taking the square root of a number are standard procedures in statistics. In both cases the mathematical theory involves raising a number to a power, 2 for squaring a number and ½ for taking the square root. The powers 2 and ½ are called *exponents*, while the number being operated on is called the *base*. Thus, this section will review the basic properties of exponents and their use with the familiar square root or radical ($\sqrt{}$) symbol.

4.1 Exponents

When we write a number such as 2, 7, or 32, it is understood that the implied exponent of each number is 1; that is, $2 = 2^1$, $7 = 7^1$, and $32 = 32^1$. Consequently, we rarely even bother to write the exponent of 1. However, when we repeatedly multiply a number times itself, we can write the operation using exponents. For example,

$$2 \cdot 2 \cdot 2 \cdot 2 \cdot 2 = 2^5 = 32$$

In this case, we raised the base number 2 to the exponent 5 and obtained 32.

Exponents follow certain rules, listed below in the box. In each rule, the letters a and b symbolize the base (like 2 in the previous example), and the letters m and n represent exponents (like 5 in the previous example).

RULES
WORKING WITH EXPONENTS

1. $a^m \cdot a^n = a^{m+n}$

2. $(a^m)^n = a^{mn}$

3. $\dfrac{a^m}{a^n} = a^{m-n}$, *provided a is not the number 0.*

4. $(ab)^m = a^m b^m$

5. $\left(\dfrac{a}{b}\right)^m = \dfrac{a^m}{b^m}$

6. $a^0 = 1$, *provided a is not the number 0.*

7. $\dfrac{1}{a^n} = a^{-n}$, *provided a is not the number 0.*

To illustrate each rule, consider the following examples.

1. $2^3 \cdot 2^2 = 2^{3+2} = 2^5 = 32$

2. $(2^3)^2 = 2^{3 \text{ times } 2} = 2^6 = 64$

3. $\dfrac{2^5}{2^3} = 2^{5-3} = 2^2 = 4$

4. $(2 \cdot 3)^2 = 2^2 \cdot 3^2 = 4 \cdot 9 = 36$

5. $(\dfrac{6}{10})^2 = \dfrac{6^2}{10^2} = \dfrac{36}{100} = .36$

SUMMARY EXAMPLE
Exponents

Express the following as x to a single exponent:

(1) $x^4 x^3$ (2) $(x^2)^4$ (3) $\dfrac{x^4}{x^6}$

Part (1)	Rule 1 - add the exponents	$x^4 x^3 = x^{4+3} = x^7$ **Answer** ↗
Part (2)	Rule 2 - multiply the exponents	$(x^2)^4 = x^{2 \text{ times } 4} = x^8$ **Answer** ↗
Part (3)	Rule 3 - subtract the exponents	$\dfrac{x^4}{x^6} = x^{4-6} = x^{-2}$ **Answer** ↗

Exponents and Radicals

 ▷▶USING YOUR TI-83 ▷▶

We will replicate the previous example on exponents, but we will have to assign a numerical value to x in order to do so.

〜〜〜〜〜〜〜〜〜〜〜〜〜〜〜〜〜〜〜〜〜〜〜〜〜〜〜〜

USE THE TI-83   Letting x = 2, solve the following:

(1) x^4x^3 (2) $(x^2)^4$ (3) $\dfrac{x^4}{x^6}$

---

(1)   ▷▶   *Press 2, ∧ (power key, below CLEAR), 4, X, 2, ∧, 3, ENTER.*

DISPLAY: **128, which is $2^7 = x^7$**

---

(2)   ▷▶   *Press 2, ∧, 2, ∧, 4, ENTER (or 2, x^2, ∧, 4, ENTER).*

DISPLAY: **256, which is $2^8 = x^8$**

---

(3)   ▷▶   *Press 2, ∧, 4, division key, 2, ∧, 6, ENTER*

DISPLAY: **.25, which is $2^{-2} = x^{-2}$**

(Press MATH, 1, ENTER to convert the result to a fraction.)

〜〜〜〜〜〜〜〜〜〜〜〜〜〜〜〜〜〜〜〜〜〜〜〜〜〜〜〜

The third rule has at least one interesting implication. Suppose we wrote:

$$\frac{3}{3}$$

Obviously, three divided by three is 1, but according to the third rule, we get:

$$\frac{3}{3} = \frac{3^1}{3^1} = 3^{1-1} = 3^0$$

It must be the case that $3^0 = 1$, from which we can generalize that <u>any number raised to the 0 exponent is 1</u>. The only exception to this rule is the special case of 0^0, which we will not consider. However, it is appropriate to point out a consequence that we sometimes forget: <u>Dividing by 0 is not permitted</u>. If you forget, your calculator will remind you: Turn on your calculator and try 5 ÷ 0. What does the display show?

Exponents and Radicals   81

The third rule also enables us to express a number raised to a negative exponent. For instance, the fraction $^9/_{27}$ can be reduced to $^1/_3$, and $^1/_3$ can be expressed as 3^{-1}:

$$\frac{9}{27} = \frac{3^2}{3^3} = 3^{2-3} = 3^{-1} = \frac{1}{3}$$

These last two results ($3^0 = 1$; $^1/_3 = 3^{-1}$) are included as rules 6 and 7 in the previous box of rules for exponents.

4.2 Fractional Exponents

So far we have considered only whole numbers as exponents. Now suppose that the exponent is a fraction such as $½$, $^1/_3$, and so on.

Recall that the number 5 is the square root of 25 since $5^2 = 25$. In general, a number, r, is a *square root* of a positive number, a, if $r^2 = a$. An alternative way of writing $r^2 = a$ is to use fractional exponents: $r = a^½$. Further, we can write $a^½$ using the *radical symbol* ($\sqrt{}$):

$$a^½ = \sqrt{a}$$

Note that –5 also is a square root of 25 since $(-5)^2 = 25$. Consequently, when we reverse the process and write $\sqrt{25}$, we should realize that there are always 2 roots, a positive one and a negative one. In statistics, we will be interested in the positive root mostly, so for the remainder of this book we will use the positive root only.

In a like manner, we say that the number r is an *nth root* of a positive number, a, if $r^n = a$, or if $r = a^{1/n}$. In this case we write:

$$a^{\frac{1}{n}} = \sqrt[n]{a}$$

The extension to an exponent like $^m/_n$ follows if we view $^m/_n$ as m times $^1/_n$.

Thus, when we write $a^{\frac{m}{n}}$, we should view this as $(a^m)^{\frac{1}{n}}$; hence,

$$a^{\frac{m}{n}} = \sqrt[n]{a^m}$$

Let us summarize these results on fractional exponents in the following box.

Exponents and Radicals

RULES

FRACTIONAL EXPONENTS

1. $a^{\frac{1}{2}} = \sqrt{a}$

2. $a^{\frac{1}{n}} = \sqrt[n]{a}$, *provided a is not a negative number.*

3. $a^{\frac{m}{n}} = \sqrt[n]{a^m}$, *provided a is not a negative number.*

*In Rules #2 and #3, it is possible to find the nth root of a negative number, a, provided n is an odd-numbered root. For example,

$$\sqrt[3]{-8} = -2$$

For the exercises in our book we will work with only positive numbers, a. Thus, we defined Rules #2 and #3 (above) in a more restrictive sense than other math books.

		SUMMARY EXAMPLE Fractional Exponents	
		Express the following using the radical symbol ($\sqrt{\ }$): $(1)\ x^{\frac{1}{2}}\qquad (2)\ x^{\frac{1}{5}}\qquad (3)\ x^{\frac{3}{8}}$	
Part (1)	Rule 1	$x^{\frac{1}{2}} = \sqrt{x}$	←**Answer**
Part (2)	Rule 2	$x^{\frac{1}{5}} = \sqrt[5]{x}$	←**Answer**
Part (3)	Rule 3	$x^{\frac{3}{8}} = \sqrt[8]{x^3}$	←**Answer**

 ## ▷▶USING YOUR TI-83 ▷▶

The TI-83 can evaluate fractional exponents, but as in the previous example, we will need to assign a numerical value to x for illustrative purposes. We will also find it simplest to enter fractional exponents as decimal numbers.

~~~~~~

USE THE TI-83   Letting x = 25, solve the following:

(1) $x^{\frac{1}{2}}$ (2) $x^{\frac{1}{5}}$ (3) $x^{\frac{3}{8}}$

(1) ▷▶   Press 25, ∧, .5, ENTER.   (.5 being ½ in decimal form)

[D][I][S][P][L][A][Y]: $5 = 25^{1/2} = \sqrt{25}$ (or, press 2nd, $\sqrt{\ }$, 25, ENTER)

(2) ▷▶   Press 25, ∧, .2, ENTER.

[D][I][S][P][L][A][Y]: 1.90 (to 2 decimal places) $= 25^{1/5} = \sqrt[5]{25}$

(3) ▷▶   Press 25, ∧, .375, ENTER.

[D][I][S][P][L][A][Y]: 3.34 (to 2 decimal places) $= 25^{3/8} = \sqrt[8]{25^3}$

~~~~~~

4.3 Using a Less Sophisticated Calculator

Actually raising a number to a power or finding a root of a number is best done with a calculator having a $\sqrt{\ }$ key or a y^x, x^y, $y^{1/x}$, or $x^{1/y}$ key. If you have a basic calculator that only adds, subtracts, multiplies, and divides, then you will not be able to perform the operations we are about to describe. At a minimum your calculator should have a key labeled x^y (or y^x). Such a key will likely be found on the field of keys that are <u>above</u> the number keys 1 through 9 plus 0. If you have such a key, continue reading as we explain how to use it. However, because of the variety of calculators on the market, the keystrokes that we recommend may not work for your calculator. Though we will do the best we can to explain a proper key sequence, you always have the option of referring to your calculator's user manual.

For example, suppose your calculator has a key labeled x^y and you wish to find 6^3. To use your calculator to find the answer, press the key labeled 6, then the x^y key, then the key labeled 3, and then the key labeled =. Your calculator display should show the number 216. Note that you should press these keys <u>sequentially</u>, not simultaneously.

Look very closely at your calculator keys again, especially right above the x^y key. Try to find the symbol $x^{1/y}$ (or $y^{1/x}$). If you have the x^y key, then likely the $x^{1/y}$ symbol is stamped on the calculator faceplate immediately above the x^y key.

Next, look in the upper right-hand or left-hand corner of the field of keys for one that is labeled 2nd, or SHIFT, or INV. (Or, if the $x^{1/y}$ symbol is color-coded, such as in blue or yellow, then look for a blue or yellow key.)

To access the $x^{1/y}$ operation we will first have to press the 2nd, or SHIFT, or INV, or color-coded key and then the x^y key. As an analogy, think about the keys on a typewriter or keypad for a computer. Pressing the "A" key gives the lower case a, while pressing the Shift key plus the "A" key gives the upper case A. Similarly, pressing the x^y key allows us to compute numbers like 6^3, while pressing the 2nd, SHIFT, INV, or color-coded key and then pressing the x^y key allows us to compute numbers like $6^{1/3}$. In a sequence of keystrokes such as the one we just described, please remember to <u>release</u> each key before you press the next key.

For example, let us find $6^{1/3}$, and let us assume the $x^{1/y}$ symbol is immediately above the x^y key and that the access to this operation is from a SHIFT key. In sequence, press the key labeled 6, (release), then the SHIFT key, (release), then the x^y key, (release), then the key labeled 3, (release), and finally the key labeled =. Your calculator display should show the number 1.817120593. In summary, we would write our answers as:

$$6^3 = 216$$
$$6^{1/3} = 1.817120593$$

(Note: In the last example, if your calculator displays 1.8 or 1.82, then you may need to adjust the number of decimal places that are shown. Refer to your calculator manual for directions about making this adjustment.) For practice using your calculator, refer to Exercise 4.11, which lists the correct answers to problems involving fractional exponents.

Exponents and Radicals

SUMMARY EXAMPLE
Using a Calculator
(with x^y and SHIFT keys)

Use the x^y key to find: (1) $(4.51)^3$ (2) $(6.2)^{1/5}$

Part (1)	Enter 4.51; Press x^y key; Enter 3; Press = key	$(4.51)^3$ = **91.733851** ← Answer
Part (2)	Enter 6.2; Press SHIFT key; Press x^y key; Enter 5; Press = key	$(6.2)^{1/5}$ = **1.440384164** ← Answer

The use of the SHIFT key (followed by the x^y key) in the previous Summary Example presumes that the function $x^{1/y}$ is above the x^y key.

USING EXCEL

We can use Excel to raise numbers to powers that are whole numbers as well as fractional values. Make any cell the active cell and click on the Paste Function key, f_x. Highlight All in the Function Category column and scroll down the Function Name column until you find the function labeled POWER. Highlight POWER and click on the word OK near the lower right hand corner of the window.

If we wanted to evaluate $(4.51)^3$, for example, we would key in the value 4.51 in the Number box, press the **Tab** key, and then enter the number 3 in the Power box. When we click on the word OK near the lower right hand corner of the POWER window, the number 91.733851 appears in the active cell of the Excel spreadsheet.

Exponents and Radicals

EXCEL F.Y.I.

To raise a number to a power ...

    (1) ... make any cell the active cell,

    (2) click on the Paste Function key (f_x),

    (3) highlight All in the Function Category list,

    (4) highlight POWER in the Function Name list,

    (5) click on the word OK,

    (6) key in the value of the number in the Number box,

    (7) press the **Tab** key,

    (8) key in the value of the power in the Power box,

    (9) click on the word OK.

4.4 Applications

We will spotlight four settings in statistics where knowledge of exponents is needed. The first application occurs in descriptive statistics when studying *measures of dispersion* (also called *measures of spread*). Suppose we had three columns of data labeled f, M, and fM, as indicated below.

f	M	fM
4	3	12
5	7	35
8	11	88
3	15	45

Further suppose we need this total: ΣfM^2. Although it may be tempting either to square the four numbers in the fM column, and then add, or to add the four numbers in the fM column, and then square, both tactics are incorrect. The third rule of exponents reveals why. For if we square fM, we get f^2M^2: $(fM)^2 = f^2M^2$. The resulting total of numbers would produce Σf^2M^2, not ΣfM^2 as desired.

There are several ways to generate ΣfM^2 from the given columns of information. One way would be to generate an M^2 column by squaring each of the M-values, and then generate an fM^2 column by multiplying each f by each M^2. Of course, then we would total the fM^2 column to get ΣfM^2. An alternative way is to use the first law of exponents by recognizing that $fM^2 = M(fM)$. Thus, we can multiply each M by the corresponding fM:

f	M	fM	fM²
4	3	12	36
5	7	35	245
8	11	88	968
3	15	45	675

$$1924 = \Sigma fM^2$$

Our second application comes from probability material on the *binomial probability distribution*. Here, we will be expected to raise decimal fractions to integer exponents. For example, one such computation might be $10(.5)^2(.5)^3$. By the first law of exponents we can add the exponents and write $10(.5)^2(.5)^3 = 10(.5)^5$. Then we can use the x^y key on our calculator to find $(.5)^5$: Enter .5; press the x^y key, enter 5, and then press the = key to get .03125. Then multiply this number by 10 to yield .3125. In other words, $10(.5)^2(.5)^3 = .3125$.

As another example, suppose we had to evaluate $(.2)^0(.8)^4$. Since any number raised to the 0 exponent is 1, this computation reduces to $(.2)^0 (.8)^4 = 1(.8)^4 = .4096$.

Here is one for you to try: $10(.3)^2(.7)^3$. "Chain" the steps together with your calculator by translating the following phrase into keystrokes: 10 times (.3) squared times (.7) to the 3rd power equals _____. (The answer is on the next page at the end of this section.)

The third illustration of exponents features the computation of a descriptive statistic called the *coefficient of skewness*. We will need to raise a number to the $^3/_2$ exponent, such as

$$(7.51)^{\frac{3}{2}}$$

From Fractional Exponents Rule #3 we know that

$$(7.51)^{\frac{3}{2}} = \sqrt{(7.51)^3}$$

First, we use the x^y key on our calculator to find $(7.51)^3 = 423.564751$. Then we take the square root:

$$\sqrt{423.564751} = 20.5806...$$

Exponents and Radicals

 **USING EXCEL**

Computing the coefficient of skewness using Excel is an application of raising a number to a power, which we illustrated in section 4.3. The difference in this section is that the power is fractional whereas the example we previously illustrated involved raising the number 4.51 to a whole number power (3).

Make any cell the active cell and click on the Paste Function key, f_x. Highlight All in the Function Category column and POWER in the Function Name category. Then click on the word OK. To compute 7.51 raised to the 3/2 power, enter 7.51 in the Number box, press the **Tab** key, and then enter the symbol "3/2" in the Power box. Notice to the right of the Power box that 3/2 was interpreted as the number 1.5. (We could have keyed in "1.5" rather than 3/2, but Excel will do the arithmetic for us if we choose to enter fractional values in lieu of decimals.) Click on OK and Excel returns the number 20.5806888 (perhaps rounded to fewer places) in the active cell.

Our last example deals with the *geometric mean*, which is another descriptive statistic. Again we will be faced with the problem of raising a number to a fractional power. For instance, we might have to evaluate the fifth root of 1.7125:

$$\sqrt[5]{1.7125} = (1.7125)^{\frac{1}{5}}$$

To find this number we use the SHIFT and x^y keys: enter 1.7125, press the SHIFT key, enter 5, press the x^y key, and finally press the = key to get 1.113592035.

(Answer to problem on previous page: 10 times (.3) squared times (.7) to the 3rd power equals .3087.)

4.5 Scientific Notation

In the process of raising a number to an exponent using our calculators, we may exceed the number of display positions on the calculator's screen. Most calculators allow seven to ten digits to appear at one time across the face of the screen. (To determine how many digits your calculator displays, simply start entering the counting numbers, beginning with 1, until the display will not accept any more digits.)

Let us suppose that our calculator displays a maximum of eight digits. If we wished to multiply 360,360 times 3,628,800, then we should get 1,307,674,368,000. Unfortunately, this number involves 13 digits, which will not fit on our 8-digit-display calculator. So, when we attempt to multiply 360,360 and 3,628,800, our calculator automatically switches into scientific notation to display the answer.

Scientific notation is a method of representing any number as the product of a number between 1 and 10 and a power of 10. The power of ten is the number of digits through which we must move the decimal point in order to produce a number between 1 and 10. For instance, the number 1,307,674,368,000 can be expressed in scientific notation as $1.307674368000 \times 10^{12}$. The number 1.307674368000 is clearly a number between 1 and 10, and the power of 12 means we must move the decimal point to the right through 12 digits in order to restore the original number, 1,307,674,368,000.

Our calculator still cannot display all the digits, so only the first 7 digits (rounded off) to the right of the decimal are shown. Also, most calculators cannot display the "$\times 10^{12}$" part of the scientific notation, so only the exponent 12 appears, usually in the far right corner of the screen and perhaps as a smaller sized number almost appearing to be an exponent. For example, 1.3076744×10^{12} might appear as either 1.3076744 12 or 1.3076744 ^{12} on your calculator display. Another possible display on some calculators might be 1.3076744E12. The "E" is used as an abbreviation for "exponent" and, in this case, signifies that the scientific notation exponent is 12, which is the number that follows the E in the calculator display.

Here are several examples of how scientific notation can be used to represent any number:

8	=	8.0×10^0
12	=	1.2×10^1
949	=	9.49×10^2
1,593	=	1.593×10^3
6,250,000	=	6.25×10^6

Scientific notation also works for very small decimal fractions. For instance, .000188 can be written as 1.88×10^{-4}. Notice that the exponent is negative 4 in this case since we must move the decimal point in 1.88 four places to the left to restore the original number, .000188. Here are other examples:

.7	=	7.0×10^{-1}
.562	=	5.62×10^{-1}
.0000041	=	4.1×10^{-6}

Scientific notation is more likely to appear on your calculator than it is in the textbook. Nevertheless as you use your calculator to solve problems, there may be occasions when you discover that an intermediate calculation has become very large or very small, causing the calculator to resort to scientific notation.

Some calculators do not have scientific notation capability. If a number beyond its capacity is asked for, it is likely either to "lock up" or display an error message. Calculators with scientific notation often cannot display powers beyond the 99th.

Exponents and Radicals

SUMMARY EXAMPLE
Scientific Notation

(1) Write 18,306 in scientific notation
(2) Write out 7.33 X 10^{-3}, eliminating scientific notation

Part (1)	Move decimal point 4 places	$18,306 = 1.8306 \times 10^4$ ← **Answer**
Part (2)	Move decimal point 3 places	$7.33 \times 10^{-3} = .00733$ ← **Answer**

 ▷►**USING YOUR TI-83** ▷►

The TI-83 will allow up to ten digits on screen before it will convert to scientific notation. It will also resort to scientific notation if a number's absolute value is less than .001.

USE THE TI-83

If you press MODE, you will see three options on the first line. Usually you will want to be in *Normal* ; we'll switch to *Sci* to do this: display 18,306 in scientific notation.

▷►   *Input 18306, MODE, right arrow, ENTER, CLEAR, ENTER.*

🄳🄸🅂🄿🄻🄰🅈: **1.8306 E4, which is equivalent to 1.8306 X 10^4 as in the hand-solved example above. Or, you can interpret the 4 to mean: Move the decimal 4 places to the right to get out of scientific notation. Now, since you likely do not want to remain in *Sci* mode, let's press MODE, left arrow, ENTER, CLEAR, ENTER to restore the original 18,306.**

Chapter 4 Exercises

4.1 Express the following as x to a single exponent:

    (a) x^5x^6

    (b) $(x^3)(x^2)(x^5)$

    (c) $(x^{1/2})^4$

    (d) $(x^3)^8$

    (e) $\dfrac{x^7}{x^4}$

    (f) $\dfrac{x^{-2}}{x^{-3}}$

4.2 Express the following using the radical symbol ($\sqrt{\ }$):

    (a) $x^{1/3}$

    (b) $x^{2/9}$

    (c) $x^{-1/2}$

    (d) $x^{6/5}$

4.3 Write the following as x to a negative exponent:

    (a) $\dfrac{1}{x^2}$

    (b) $\dfrac{1}{x^3}$

    (c) $\dfrac{1}{\sqrt{x}}$

4.4 Express the following using the radical symbol ($\sqrt{\ }$):

    (a)   the cube root of 7

    (b)   the fifth root of r^2

    (c)   the nth root of 21

4.5 Refer to Exercise 4.4. Use your calculator to find the indicated quantities. For part b, let r = 4. For part c, let n = 8. Round off your answers to three decimal places.

continued on the next page ...

Exponents and Radicals

Chapter 4 Exercises (continued)

4.6   Use your calculator to find the indicated quantities. Round off your answers to four decimal places, as needed.

(a)	8^5	**(d)**	$(8.56)^3$
(b)	$15\,(.3)^6\,(.7)^0$	**(e)**	$(6.99)^{1/3}$
(c)	$\sqrt[5]{1.0259}$	**(f)**	$(4.18)^{3/2}$

4.7   Express the following numbers in scientific notation:

(a)	79268	**(d)**	.6561
(b)	890.3274	**(e)**	.000000009426
(c)	$(12345)^2$	**(f)**	1

4.8   Rewrite the following numbers, without the scientific notation:

(a) 8.8×10^{-3}
(b) 1.095×10^5
(c) 7.33×10^{-6}
(d) 2.15423×10^3
(e)   3.3431 05 (calculator display)
(f)   1.297 11 (calculator display)

4.9   Find $\Sigma f M^2$ and $\Sigma f(M - 8)^2$ from the following table:

f	M
2	2
4	7
4	12

4.10   Find $\Sigma x^2 P(X = x)$ from the following information:

x	P(X = x)
1	.4
2	.1
3	.2
4	.3

continued on the next page ...

Chapter 4 Exercises (continued)

4.11 (Practice using your calculator) Verify the following results using the x^y, y^x, $x^{1/y}$, or $y^{1/x}$ key. Note that your calculator may not be able to produce as many decimal places as these results indicate.

(a) $\sqrt{23.6} = 4.857983121$

(b) $7.2^5 = 19{,}349.17632$

(c) $12^{1/6} = 1.513085749$

(d) $8^{4/5} = 5.278031643$

POSTTEST #4
Allow 15 Minutes
Calculator with an x^y key needed
Each answer worth 1 point -- maximum score 12

4.1 Write the following as x to a single exponent:

(a) $x^5(x^5)$

(b) $(x^5)^5$

(c) $\dfrac{x^{-5}}{x^5}$

4.2 Express the following using the radical symbol ($\sqrt{}$):

(a) the fifth root of 8.

(b) the seventh root of x raised to the 3rd power.

(c) $x^{-1/3}$

4.3 Write out the following numbers, without the scientific notation:

(a) 6.56×10^{-2}

(b) 7.302×10^5

(c) 9.8288^{07} (a calculator display)

4.4 Use your calculator to find the following numbers; round-off to 5 decimal places, as needed.

(a) $(3.2)^5$

(b) $3^{5/6}$

4.5 If it is possible, express X^0 as a number.

Exponents and Radicals

NOTE TO STUDENTS

The probability material in Chapters 5, 6, and 7 may not be needed for some statistics courses or may be covered in detail for other statistics courses. Please check with your instructor to determine the level at which probability will be covered. Also, unless your school has a finite math prerequisite (where the material in these three chapters would be found), you may not have been exposed to this material in your prior math courses.

PRETEST #5
Allow 20 Minutes
1 point each part (Question 1);
1 point each thereafter -- maximum score 10
(a calculator is recommended)

5.1   Evaluate the following:

    (a) 6!   **(b)** $_4C_2$ **(c)** $_8P_3$

    Note in part b that the C-notation refers to "combination" while in part c the P-notation indicates "permutation."

5.2   For which one of the counting rules -- permutations rule, combinations rule, or product rule -- is order important?

5.3   One of the games that a state offers through its state lottery is Pick 3. To play, a player must select 3 digits from among one through nine plus zero. A selected digit can be repeated. How many different 3-digit numbers are possible?

5.4   Another popular lottery game involves selecting any 5 numbers from the numbers 1 through 35. Once selected, a number cannot be repeated. How many different groups of 5 number selections are possible?

continued on the next page ..

5.5 Refer to Question 5.3. A popular bet in the Pick 3 game is to "box" the three selected digits. Boxing a set of digits means that you buy a ticket for all the different ways that the digits can be rearranged. Assuming the three selected digits are all different (no repeats), how many tickets would you have to buy to box the digits?

5.6 A committee of 8 people must elect a chairperson and a secretary. How many different ways can these two positions be filled within the committee?

5.7 A grocery store plans to upgrade 3 of its regular checkout lines to accommodate purchases made with a credit card. If there are 12 regular checkout lines (excluding the express lines), in how many different ways could the 3 line upgrades occur?

5.8 A recovering heart attack patient was told to reduce the fat in his diet by eating more fruits, vegetables, and whole grain baked products. He was given a list of "heart healthy" foods including 12 fruits, 8 vegetables, and 3 baked products. If he selects one from each category per day, how many days will elapse before he repeats a fruit-vegetable-baked product set?

Combinations and permutations are two of the most well-known counting rules that are used in probability. We will discuss both of these rules, as well as a third rule, which is not as well-known and does not have a universally agreed upon name.

Oftentimes we enumerate all the possible outcomes of an activity. For example, a football coach once remarked that there were only three things that can happen when you attempt to pass the ball on offense and two of them are bad. He was referring to these outcomes: (1) the pass is incomplete; (2) the pass is intercepted by the defense; or (3) the pass is complete.

The ability to know or to be able to count all the outcomes is crucial in solving subsequent probability problems as well as everyday problems. For instance, the telephone company needs to know how many different telephone numbers are possible, or a state needs to know how many license plates can be issued that consist of certain combinations of letters and numbers.

The enumeration of all the possible outcomes for these type of problems is not as quick and simple as enumerating the outcomes for the football coach. As a result we need to learn how to use some mathematical counting rules to answer the question "how many?" In real life, in games of chance, and in most probability problems, the rules we present here will be sufficient. However, there are *many* other rules in the field of combinatorics that we cannot cover.

5.1 Factorial Notation

In Chapter 3 we introduced the Greek letter sigma (Σ) as a symbol to represent the operation of addition. A symbol also is used to represent the multiplication of successive integers from 1 through n. The symbol is called *factorial notation* and is denoted by an exclamation point (!) following the letter n, as in "*n!*"

We read the symbol *n!* as "n – factorial" and simply multiply the first n integers together in any order to evaluate it:

$$n! = 1 \cdot 2 \cdot 3 \cdot 4 \cdot 5 \cdots (n - 1) \cdot n = n \cdot (n - 1) \cdots 5 \cdot 4 \cdot 3 \cdot 2 \cdot 1$$

For example,

$$3! = 1 \cdot 2 \cdot 3 = 3 \cdot 2 \cdot 1 = 6$$

$$7! = 1 \cdot 2 \cdot 3 \cdot 4 \cdot 5 \cdot 6 \cdot 7 = 5040$$

There are two integer values -- 0 and 1 -- for which we must make special definitions for their factorials. In both cases, we assign the value 1 to the symbols 0! and 1!:

$$0! = 1 \qquad 1! = 1$$

For small numbers like 3 and 7 we can find 3! and 7! by hand, but as the number gets bigger, the corresponding factorial is more difficult to find. Your calculator can be used to do the successive multiplications, or you may have a sufficiently sophisticated calculator that has a factorial key. Look over the keyboard for a key or symbol labeled n! or x!.

On our inexpensive calculator the factorial function is indicated with the x! symbol located above the $^1/_x$ key. Hence, to access the factorial function we first have to press the SHIFT key. If your calculator has a factorial key, then you would push it directly to access it.

For example, to find 10! on our calculator, we enter 10, press the SHIFT key, then press the $^1/_x$ key (to access the factorial function above the key), and our calculator automatically displays the number 3,628,800.

Depending on the number of digits in your calculator display, the factorials will not "overflow" the display until about 14! Below 14!, the entire number appears in the display; for 14! and larger numbers, the factorial appears in scientific notation. For instance, 13! appears as 6,227,020,800, but 14! yields the display 8.71782912^{10}, which means $8.71782912 \times 10^{10}$ or 87,178,291,200.

Most calculators cannot do large factorials. With our calculator we are limited to 65! Beyond this number, our calculator displays a couple of dashes to indicate that we have exceeded its capacity. You might want to try a few large factorials in your calculator to determine its capacity. Some accuracy is lost for large factorials, because only the first 8 or so digits of the number are displayed. The remaining digits simply won't fit on the calculator display so they are rounded off, which introduces some inaccuracy. For instance, the correct value of 16! and our calculator display of 16! are shown below:

$$16! = 20,922,789,888,000 \text{ (correct)}$$

$$16! = 2.092278989 \times 10^{13} = 20,922,789,890,000 \text{ (calculator)}$$

To fathom how incredibly large factorials become, consider this: Suppose a creature in a parallel universe tries to perform all the rearrangements of an encyclopedia set numbered from 1 through 20. For example, one such arrangement (from left to right) is 1-2-3-4 and so on, up to 20. A second such arrangement is 2-1-3-4 and so on, up to 20. The number of different such arrangements is 20! If the creature worked at the rate of one new arrangement per second and if the creature began this process at the same time that the Big Bang started our universe, the alien would not yet be finished.

	SUMMARY EXAMPLE Factorial Notation	
Compute	**(1)**	**5! (by hand)**
	(2)	**15! (using "an ordinary" calculator factorial key)**
Part (1)	Multiply the first 5 integers	$5! = 1 \cdot 2 \cdot 3 \cdot 4 \cdot 5 = \textbf{120}$ ⬅ **Answer**
Part (2)	Enter 15; press the factorial key	$15! = \textbf{1.307674368}^{12}$ ⬆ **Answer** ⬇ or **1,307,674,368,000**

▷▶ USING YOUR TI-83 ▷▶

Factorials are quickly evaluated on the TI-83. Below we will repeat the computations performed in the previous example.

~~~~~~

USE THE TI-83   **Evaluate**   (1)   **5!**
                                        (2)   **15!**

___

▷▶   *Press 5, MATH, left arrow, 4, ENTER*

D I S P L A Y: **120**

___

▷▶   *Press 15, MATH, left arrow, 4, ENTER*

D I S P L A Y: **1.307674368E12**

(Scientific notation is used by the TI-83 since the result is more than 10 digits.)

~~~~~~

USING EXCEL

There is a built-in function in Excel called FACT to compute factorials. To use this function, make any empty cell the active cell and click on the Paste Function key, f_x. Highlight the category All in the Function Category column on the left and scroll down the Function Name column until you find FACT. Highlight FACT and click on the word OK in the lower right corner.

Simply key in the number for which you desire the factorial in the Number window of the FACT screen and click on OK near the lower right corner. The result appears in the active cell you have selected.

For example, 14! appears as the number 87178291200 (no commas). Numbers above 14 will return factorials written in scientific notation. For instance, the factorial of 15 appears as 15! = 1.30767E+12. The largest number that Excel can handle is 170!.

EXCEL F.Y.I.

To find the factorial of a number ...

        (1) ... make an empty cell active,

        (2) click on the Paste Function key (f_x),

        (3) highlight All in the Function Category list,

        (4) highlight FACT in the Function Name list,

        (5) click on the word OK,

        (6) key in the appropriate number,

        (7) click on the word OK.

5.2 Permutations Rule

The permutations rule is the first of the three counting rules that we will consider. Each counting rule is based on the concept of items occupying positions or slots in a sequence. A *permutation* is an ordered sequence of items; the <u>order</u> in which the items are lined up from left-to-right is <u>important</u>. For instance, suppose you plan to eat a carton of yogurt every day at work for lunch, so you buy five different cartons of yogurt each week at the grocery store. To understand the counting rules, we have to distinguish the items from the slots. In this example think of the slots as each day of the week -- Monday through Friday. There are five slots. Think of the items as the different cartons of yogurt; perhaps imagine five different flavors such as apple, banana, peach, strawberry, and cherry. There are five items.

If we ate the apple (A) yogurt on Monday, banana (B) on Tuesday, peach (P) on Wednesday, strawberry (S) on Thursday, and cherry (C) on Friday, then the letters ABPSC describe the sequence in which we consumed our cartons of yogurt this week. But if we decided to switch and eat the cherry yogurt on Thursday and the strawberry yogurt on Friday, then the sequence becomes ABPCS and is different from the first sequence.

The permutations rule is used to tell us the number of different sequences that are possible when we rearrange the items in the slots. In the yogurt example the permutations rule would answer the question, "How many different sequences of yogurt are possible throughout the week?" To answer this specific question we need to first define general terms and symbols associated with the permutations rule:

$$n \quad = \quad \text{number of items}$$
$$r \quad = \quad \text{number of slots to be filled}$$
$$_nP_r \quad = \quad \text{number of different sequences of length r}$$

The symbol $_nP_r$ is also written as:

$$P_r^{\,n}$$

in some texts, and it is also referred to as "the number of permutations of n things taken r at a time."

In the yogurt example, n = 5 items (yogurt cartons), r = 5 slots (days of the week), and $_5P_5$ is the symbol for the number of different sequences of yogurt. (Sometimes a "sequence" will be referred to as a "permutation." The question, "How many different *sequences* of yogurt are possible?" is synonymous with the question, "How many different *permutations* are possible?") The following formula tells us how to attach a number to the symbol $_nP_r$:

∿∿ ∿∿ ∿∿ ∿∿ ∿∿ ∿∿ ∿∿ ∿∿ ∿∿ ∿∿ ∿∿ ∿∿ ∿∿ ∿∿ ∿∿ ∿∿ ∿∿ ∿∿ ∿∿

《FORMULA》

PERMUTATIONS RULE

$$_nP_r \quad = \quad \frac{n!}{(n-r)!}$$

∿∿ ∿∿ ∿∿ ∿∿ ∿∿ ∿∿ ∿∿ ∿∿ ∿∿ ∿∿ ∿∿ ∿∿ ∿∿ ∿∿ ∿∿ ∿∿ ∿∿ ∿∿ ∿∿

To complete our yogurt example, the number of different sequences of n = 5 yogurt cartons and r = 5 days is:

$$_5P_5 \quad = \quad \frac{5!}{(5-5)!} = \frac{5!}{0!} = \frac{1 \cdot 2 \cdot 3 \cdot 4 \cdot 5}{1} = 120$$

There are 120 different sequences of the letters A, B, P, S, and C.

SUMMARY EXAMPLE
Permutations Rule

In the preliminary round of the World Cup soccer tournament, the top 2 teams in each group of teams automatically qualify for the elimination rounds. The order of finish within each group is important because it dictates the seeding of teams in the final round. If there are 4 teams in a group during the preliminary round, how many different sequences of teams could finish first and second in the group?

Do this first:	Identify the *items* and the value of n	The *items* are the teams in the group; n = 4
Do this second:	Identify the *slots* and the value of r	The *slots* are the first and second positions in the standings; r = 2
Do this third:	Use formula for the permutations rule	$_4P_2 = \dfrac{4!}{(4-2)!}$ $= \dfrac{4!}{2!} = \dfrac{1 \cdot 2 \cdot 3 \cdot 4}{1 \cdot 2}$ $= 12 \leftarrow$ **Answer**

 **USING YOUR TI-83** ▷▶

The permutations function is on the same menu as the factorial selection.

USE THE TI-83

To answer the World Cup question above, we need to know the number of different sequences possible when 4 items exist and there are 2 slots to be occupied: $_4P_2$

▷▶ *Press 4, MATH, left arrow, 2, 2, ENTER.*   D I S P L A Y : **12**

USING EXCEL

The number of permutations is easily calculated in Excel using the PERMUT function. Make any empty cell the active cell and click on the Paste Function key, f_x. Highlight All in the Function Category column and PERMUT in the Function Name column. Then click on the word OK near the lower right hand corner of the Paste Function window.

In the Number box enter the value of n, the number of items. For example, enter the number 4. Then activate the Number_chosen dialog box by pressing the **Tab** key. Enter the value of r, the number of slots to be filled, and then click on the word OK near the lower right hand corner of the PERMUT window. For example, entering 2 in the Number_chosen box and then clicking OK produces the number 12 in the active cell.

EXCEL F.Y.I.

To compute $_n P_r$ ...

- (1) ... make an empty cell active,
- (2) click on the Paste Function key (f_x),
- (3) highlight All in the Function Category list,
- (4) highlight PERMUT in the Function Name list,
- (5) click on the word OK,
- (6) key in the value of **n**,
- (7) press the **Tab** key,
- (8) key in the value of **r**,
- (9) click on the word OK.

To show you that there are indeed 12 different sequences, we will list them. Suppose a particular group of four teams consisted of these countries: Germany (G), South Korea (K), Canada (C), and Sweden (S). The 12 different sequences of the top two finishers in this group are (first letter listed is first place finisher, second letter is second place finisher):

GK	KG	CG	SG
GC	KC	CK	SK
GS	KS	CS	SC

5.3 Combinations Rule

As the last summary example for the permutations rule showed, the <u>order</u> in which the items fill the slots is important. A first place finish means something different (in the next round of play) than a second place finish in the World Cup preliminary round. Notice however, that both first and second place finishers automatically qualify for the elimination rounds of the World Cup. If we disregard the order of finish and ask instead for the number of different pairs of teams that could automatically qualify by finishing first <u>or</u> second, then we would use the combinations rule. A *combination* is an unordered group of items; the <u>order</u> in which the items are lined up from left-to-right is <u>not important</u>.

The key difference between the permutations rule and the combinations rule is that order is important in the former case, but order is <u>not</u> important in the later case. Another way of stating this difference is to say that the slots to be filled are <u>distinguishable</u> for permutations but the slots are <u>indistinguishable</u> for combinations. Thus we think of *sequences* for permutations and *groups* for combinations. For the combinations rule the general terms and symbols are:

$$n \quad = \quad \text{number of items}$$
$$r \quad = \quad \text{number of slots to be filled}$$
$$_nC_r \quad = \quad \text{number of different groups of size } r$$

Other symbols may be used to represent a combination, which are equivalent to our symbol $_nC_r$:

$$C_r^n \qquad \binom{n}{r} \qquad C(n,r)$$

Also, the symbol $_nC_r$ is referred to as "the number of combinations of n things taken r at a time."

In the soccer example there are n = 4 items (teams), r = 2 slots (first and second position in the standings), and $_4C_2$ number of different groups (pairs of teams) that could qualify for the elimination rounds. The following formula tells us how to attach a number to the symbol $_nC_r$.

《*FORMULA*》
COMBINATIONS RULE

$$_nC_r \quad = \quad \frac{n!}{r!\,(n-r)!}$$

To complete the soccer example, the number of different groups of r = 2 teams that could finish first or second out of n = 4 teams is:

$$_4C_2 = \frac{4!}{2!(4-2)!} = \frac{4!}{2!2!} = \frac{1 \cdot 2 \cdot 3 \cdot 4}{1 \cdot 2 (1 \cdot 2)} = 6$$

Again it is possible to show you the 6 different groups using our previous example with the countries Germany (G), South Korea (K), Canada (C), and Sweden (S):

GK	GS	KS
GC	KC	CS

Because the order of finish is unimportant in this setting, we do not need to list GK <u>and</u> KG, for example. Either symbol -- KG or GK -- can be used to indicate that Germany and South Korea made the elimination round, but both symbol pairs are not needed.

SUMMARY EXAMPLE
Combinations Rule

A company plans to hire 3 new college graduates. After the initial round of interviews, 8 candidates will be invited back for a second interview. How many different groups of 3 can emerge from the second interview as the eventual new hires?

Do this first:	Identify the *items* and the value of n	The *items* are the 8 candidates invited for a second interview; n = 8
Do this second:	Identify the *slots* and the value of r	The *slots* are the 3 positions to be filled; r = 3
Do this third:	Use formula for the combinations rule	$_8C_3 = \dfrac{8!}{3!(8-3)!}$ $\dfrac{8!}{3!5!} = \dfrac{1 \cdot 2 \cdot 3 \cdot 4 \cdot 5 \cdot 6 \cdot 7 \cdot 8}{1 \cdot 2 \cdot 3 (1 \cdot 2 \cdot 3 \cdot 4 \cdot 5)}$ = 56 ← Answer

▷▶ USING YOUR TI-83 ▷▶

The combinations option on your TI-83 is right below the permutations function.

〰〰〰〰〰〰〰〰〰〰〰〰〰〰〰〰〰〰〰〰〰〰〰〰

USE THE TI-83   to see how many different groups of size 3 can be formed from 8 items: that is, evaluate $_8C_3$.

▷▶   *Press 8, MATH, left arrow, 3, 3, ENTER.*   D I S P L A Y :   **56**

〰〰〰〰〰〰〰〰〰〰〰〰〰〰〰〰〰〰〰〰〰〰〰〰

USING EXCEL

Make any cell the active cell and click on the Paste Function key, f_x. Highlight All in the Function Category column and COMBIN in the Function Name column, and then click on the word OK. Enter the value of **n**, the number of items in the Number dialog box. For example, enter the number 4. Then press the **Tab** key to move the cursor to the Number_chosen dialog box. Enter the value of **r**, the number of slots to be filled; for example, enter the number 2. Click on the word OK near the lower right hand corner of the COMBIN window and, for the previous values of n and r that we suggested, the number 6 should appear in the active cell.

EXCEL F.Y.I.

To compute $_nC_r$ ...

     (1)   ... make an empty cell active,
     (2)   click on the Paste Function key (f_x),
     (3)   highlight All in the Function Category list,
     (4)   highlight COMBIN in the Function Name list,
     (5)   click on the word OK,
     (6)   key in the value of **n**,
     (7)   press the **Tab** key,
     (8)   key in the value of **r**,
     (9)   click on the word OK.

Counting Rules

In your study of the binomial probability distribution you will encounter the combinations rule. As the term "binomial" implies, we are concerned with two possible outcomes -- heads and tails, defective and nondefective, win and lose, male and female, foreign and domestic, and so on. To be as general as possible, we usually refer to the two outcomes as "success" and "failure," with the understanding that the meaning of "success" and "failure" depends on the specific problem setting.

For instance, we might be interested in monitoring the next 5 cars that pass through the drive-through window of a fast food restaurant. Our interest is in determining whether the car is domestic (a "success") or a foreign (a "failure") vehicle. How many different ways could we observe exactly 3 successes among the next five cars? The combinations rule tells us the answer:

$$_5C_3 = \frac{5!}{3!(5-3)!} = \frac{5!}{3!2!} = \frac{1\cdot2\cdot3\cdot4\cdot5}{1\cdot2\cdot3(1\cdot2)} = 10$$

Here are the 10 combinations of 3 successes (S) and 2 failures (F):

SSSFF	SSFSF	SFSFS	FSFSS
SSFFS	SFSSF	FSSFS	
SFFSS	FSSSF		
FFSSS			

The notation SSSFF means, for instance, that the first 3 cars were domestic and the last 2 were foreign. We will leave the complete explanation of how the binomial is developed and used to your statistics text.

Some calculators have the ability to compute $_nP_r$ and $_nC_r$ directly provided you execute the proper sequence of keystrokes. Examine your calculator's keypad for either a key labeled $_nP_r$ or the symbol $_nP_r$ printed above a key. (Note: If your calculator has a $_nP_r$ key, then most likely it also has a $_nC_r$ key too.) On our calculator the symbols $_nP_r$ and $_nC_r$ appear above a couple of keys. There is not one universally accepted symbol for permutations (and for combinations), so be aware of the other symbols we have mentioned for these terms.

For us to access the permutations or combinations symbol we will have to first press our SHIFT key. If you have such symbols above keys on your calculator, look for the SHIFT, INV, 2nd, or color-coded key to access the symbols.

To find $_8P_4$, for example, on our calculator we enter the digit 8, then press the SHIFT key, then press the key above which the symbol $_nP_r$ is found, then enter the digit 4, and finally press the = key. Our calculator display shows the answer of 1,680; that is,

$$_8P_4 = 1680$$

To find $_8C_3$, for example, on our calculator we enter the digit 8, press the SHIFT key, press the key above which the symbol $_nC_r$ appears, enter the digit 3, and finally press the = key. Our display shows the answer of 56; that is:

$$_8C_3 = 56$$

5.4 Product Rule

The product rule (or the "mn rule" in some texts) represents the third and final counting rule that we will consider. It differs markedly from the other two rules because there are several <u>pools</u> of items rather than just one pool of items. The rules are similar in that we are still filling slots with items, but the product rule requires a separate pool of items for each slot.

To illustrate, let's suppose that you have a full-time job and you go to school three nights a week -- Monday, Tuesday, and Thursday. Pre-registration for the next term is about to begin and you need to work out a class schedule. Also suppose that classes meet one night a week, so you need to find a class for each of the three nights you can attend school. As you know, devising a class schedule is subject to many constraints such as classes that are required for your degree program, classes that are scheduled on the right nights, and classes that are open when you register!

For this example assume that there are 2 classes available for you on Monday night, 3 classes on Tuesday night, and only 1 class on Thursday night. If you take 3 classes, one on each night, how many different sets of 3 classes are possible?

To answer this question we begin by identifying slots to be filled and pools of items for each slot. The slots are the three nights when you plan to take classes. The items to fill the slots are the classes for which you will register. Notice however, that each slot has a different pool of items. On Monday night there are 2 items in the pool. On Tuesday night there are 3 items in the pool and this Tuesday night pool is different from the Monday night pool. Similarly, the Thursday night pool has only 1 item and is different from the other two pools. For the product rule there is no symbol, but there are general terms:

$$r \quad = \quad \text{number of slots to be filled}$$
$$n_1 \quad = \quad \text{number of items in the 1st pool}$$
$$n_2 \quad = \quad \text{number of items in the 2nd pool}$$
$$\vdots$$
$$n_r \quad = \quad \text{number of items in the rth pool}$$

The following formula tells us how many different sets are possible:

Counting Rules

~~~~~~

«FORMULA»
PRODUCT RULE

number of sets = n_1 times n_2 times ... times n_r

~~~~~~

To complete your class schedule example, we have r = 3 slots to be filled (1st slot, Monday; 2nd slot, Tuesday; 3rd slot, Thursday), n_1 = 2 items (classes) in the 1st pool, n_2 = 3 items (classes) in the 2nd pool, and n_3 = 1 item (class) in the 3rd pool. The number of different sets of 3 classes for which you could register is: number of sets = n_1 times n_2 times n_3 = 2 · 3 · 1 = 6

SUMMARY EXAMPLE
Product Rule

In the state of Kentucky, a truck license plate consists of two letters of the alphabet followed by a four-digit number. How many different license plates are possible?

Do this first:	Identify the *slots* and the value for r	Each position on the license plate is a slot; 2 letters + 4 digits = 6 slots; r = 6
Do this second:	Identify the *pools* for each slot and the number of items for each pool	The pools for the first 2 slots are the 26 letters of the alphabet; n_1 = 26 and n_2 = 26 The pools for the last 4 slots are the 10 digits one through nine plus zero; n_3 = 10, n_4 = 10, n_5 = 10, and n_6 = 10
Do this third:	Use product rule formula	Number of different license plates = 26 · 26 · 10 · 10 · 10 · 10 = **6,760,000 ← Answer**

5.5 Summary

Now that we have covered each counting rule, it may be helpful to combine the information about each rule into a chart. Such a chart will help us see the similarities and differences of the rules and may be useful in identifying the correct rule to use to solve a problem. The following chart contains all the relevant information about the counting rules except the formulas; they wouldn't fit on this page. Earlier in the chapter we presented each formula in a box, so refer back to these boxes as needed.

There are two ideas to keep in mind as you read this chart: the <u>task</u> you are performing and the <u>goal</u> of that task.

TASK: Fill slots with items selected from a pool (or pools).

GOAL: Count the number of different sequences, groups, or sets of items that could fill the slots.

SUMMARY OF COUNTING RULES			
CHARACTERISTICS	**PERMUTATIONS RULE**	**COMBINATIONS RULE**	**PRODUCT RULE**
1. Question to be answered	How many different sequences are possible?	How many different groups are possible?	How many different sets are possible?
2. Number of pools of items from which to select	1	1	Several
3. Number of slots to be filled	r	r	r
4. Number of items in pool (or pools)	n	n	n_1 in 1st pool n_2 in second pool $\vdots$ n_r in rth pool
5. Is order important?	yes	no	No, provided we match corresponding pools and slots
6. Assumption	After each slot is filled, the number of items in the pool decreases by one.	After each slot is filled, the number of items in the pool decreases by one.	After each slot is filled, the next pool of items is used.
7. Answer to the question	$_nP_r$	$_nC_r$	n_1 times n_2 times ... times n_r

Chapter 5 Exercises

5.1 Evaluate these factorials:

    (a)   4!   **(b)**   1!   **(c)**   8!

5.2 Use your calculator to find these factorials, if possible. Express answers in scientific notation.

    (a)   16!   **(b)**   32!   **(c)**   64!

5.3 Evaluate these expressions:

    (a) $\dfrac{10!}{8!2!}$ **(b)** $_{13}C_{10}$ **(c)** $_{9}P_{5}$

5.4 All of the following situations require that the <u>combinations rule</u> be used to count the total number of possibilities. Study the problem settings to understand why the combinations rule is appropriate and then answer the question.

    (a)   A professional baseball league in the minor leagues has 8 teams, of which the top 4 teams qualify for a post season tournament. How many different groups of 4 teams could compete in the tournament?

    (b)   A bank has 15 branches throughout the city. Three branches will be selected to receive new ATMs. How many different ways can the 3 branches be selected?

    (c)   How many different hands of 5 cards can be dealt from a deck of 52 playing cards?

    (d)   Kellogg's makes four different kinds of fruit squares cereal -- raisin, apple, strawberry, and blueberry. A local grocery store has shelf space to carry only two of the four varieties in side-by-side facings. How many different pairs of Kellogg's fruit squares cereal could be selected to fill the grocery store's shelf space?

    (e)   A bluegrass music lover has a collection of 15 different compact disks (CDS) featuring bluegrass artists. Her CD player has a tray that can be filled with 5 CDS. The CD player also has a "random" feature, which will play the tray full of CDS in a random order. How many different trays of 5 CDS could she prepare and play with the random feature?

continued on the next page ...

Chapter 5 Exercises (continued)

5.5 All of the following situations require that the <u>permutations rule</u> be used to count the total number of possibilities. Study the problem settings to understand why the permutations rule is appropriate and then answer the question. Each problem setting is an extension of the corresponding part from Exercise 5.4.

(a) In the 8-team professional baseball league the post season tournament matches the first place finisher against the fourth place finisher in a best-of-three series. The second and third place finishers compete in a separate best-of-three series. The team that finished higher in the standings retains the home field advantage in the post season matches. How many different playoff pairings of 4 teams could the league's 8 teams possibly produce?

(b) The criterion for the selection of the branch banks to receive the new ATMs is the percent increase in commercial loans relative to the previous year. The three branches with the largest percent increase receive new machines plus cash bonuses of $10,000 for the largest, $5,000 for the next largest, and $2,500 for the third largest. How many different orderings of 3 branch banks from a pool of 15 are possible?

(c) How many different ways can 5 cards be successively dealt from a 52 card deck?

(d) The grocery store manager decides to display the two Kellogg's fruit squares cereals in a stacked fashion with one variety on a shelf that is eye level for adults, and the second variety on the top shelf. (The eye level display is perceived as "better" than the top shelf display.) How many different displays of pairs of Kellogg's four fruit squares cereal are possible?

(e) The CD player has a specific order of playing CDS in the tray (unless the "random" play feature is chosen). How many different trays of 5 CDS could our bluegrass music lover arrange and play from her collection of 15 CDS if the sequence in which the CDS are played matters?

continued on the next page ...

Counting Rules

Chapter 5 Exercises (continued)

5.6 All of the following situations require that the product rule be used to count the total number of possibilities. Study the problem settings to understand why the product rule is appropriate and then answer the question.

    (a) An auto license tag (excluding vanity plates) consists of 3 letters followed by 3 digits. How many different license plates are possible?

    (b) A 3-person committee is constituted by selecting one person from each of three different departments. If the departments have 6, 8, and 10 people in them, how many different 3-person committees could be formed?

    © A stock broker recommends a portfolio mix of one stock from each of four sectors of the economy -- airlines, pharmaceutical, utilities, and precious metals. If she has identified 5 airlines, 3 pharmaceuticals, 4 utilities, and 2 precious metals stocks as attractive investments, how many different portfolios could be formed by selecting one stock from each sector?

    (d) How many 3-digit numbers can be formed from the digits 1, 3, and 5, if digits can be repeated?

    (e) A music lover has a CD collection consisting of 7 bluegrass, 8 country, 3 jazz, 4 rock 'n roll, and 5 new age CDS. If she selects one of each type to put on her CD tray, how many different collections of 5 CDS are possible?

5.7 All of the following situations refer to the corresponding parts of Exercise 5.6 and require that the product rule be used to count the total number of possibilities.

    (a) Some 3-letter combinations spell words or phrases that may be offensive to the general public. If vowels (a, e, I, o, and u) are not allowed in the second letter position, how many license plates are possible?

    (b) Suppose the chair of the committee must be a woman. If there are no women in the 10-person department and the other two departments have an equal representation of men and women, how many different 3-person committees with a female chair from the smallest department could be formed?

    (c) Suppose an investor can afford to create a portfolio with only two of the four sectors represented. One of the sectors must be the airlines sector, but the other sector doesn't matter. How many different portfolios consisting of one airlines stock and one other stock could be formed?

continued on the next page ...

Chapter 5 Exercises (continued)

(d)   How many different 3-digit numbers can be formed from the digits 0, 1, 3, and 5 if digits can be repeated?

(e)   Suppose she decides to select and play only 2 CDS, each one being a different style of music. How many different pairs of CDS can she select? (Hint: First use the combinations rule to determine the number of different pairs of music styles possible. For example, here is one such pair: bluegrass and country. Then use the product rule of each pair. Finally, add together all the products.)

5.8   In states that promote lotteries, a popular game is called Lotto in which players must select six numbers from a pool of the first N numbers. Determine the number of different 6-number combinations that are possible if N is...

(a)   40   **(b)**   42   **(c)**   44

5.9   In the Super Lotto game played simultaneously by people in several states, the pool consists of the first 52 numbers. How many 6-number combinations are possible?

5.10   Five letters are to be folded and inserted into an envelope to the correctly addressed individual. All five individuals have different names. There is only one way that the five letters can be inserted into the correct envelopes. How many incorrect ways are there?

5.11   A company plans to fill four identical managerial positions. If there are 7 equally qualified candidates, how many different sets of selections can be made?

5.12   Refer to Exercise 5.11. Suppose the company specifies that two of the positions must be filled with women. Also suppose that, of the 7 candidates, 3 are women and 4 are men.

(a)   How many different sets of 2 women can be formed from the 3 female candidates?
(b)   How many different sets of 2 men can be formed from the 4 male candidates?
(c)   How many different sets of 2 women and 2 men can be formed from the 7 candidates?

5.13   An annual competition to taste test the cabernets (a red wine) made in California results in a ranking of the top 3 wines. If 10 cabernet wines are entered into the competition, how many different orderings are possible for the top 3 positions?

5.14   Refer to Exercise 5.13. The competition also categorizes all wines into two groups: excellent and good. The top 3 wines make up the excellent category and the other 7 wines fall into the good category. How many different groupings of the 10 wines into these two categories are possible?

5.15   Prior to 1994, telephone area codes consisted of 3 digits, of which the first digit could not be the digits 0 or 1, the second digit had to be either the digit 0 or 1, and the third digit could not be the digit 0. How many different area codes were possible?

POSTTEST #5

Allow 20 Minutes
Point values -- maximum score 10
5.1 = 6, 5.2 through 5.5 = 1 point each
(A calculator is recommended)

5.1   Evaluate the following:

    (a)   4!

    (b)   0!

    (c)   20!

    (d) $\frac{7!}{2!5!}$

    (e) $_6C_3$

    (f) $_7P_6$

5.2   How many different sequences of 3 items are possible from a pool of 5 items if order is important?

5.3   A mail order company sells southwestern foods such as salsa, beans, oils, etc. One of the company's popular items is a starter basket featuring 3 different products. If the baskets are made up by selecting products from 7 different product lines, how many different starter baskets are possible?

5.4   Refer to Question 5.3. The company makes four different kinds of salsa -- mild, normal, hot, and very hot. Mild and normal salsa are sold in three different sizes -- 4 ounce, 8 ounce, and 12 ounce jars. The two hot salsas are sold in only the 4 ounce and 8 ounce jars. If a salsa sampler basket were sold that contained one of each kind of salsa (any size), how many different baskets are possible?

5.5   In a corporate downsizing, two managerial positions in a certain division have been targeted for elimination -- one position immediately and the other position six months from now. If there are six managers at risk, how many different ways could two managers lose their jobs?

CHAPTER SIX
Basic Probability

NOTE TO STUDENTS

The probability material in Chapters 5, 6, and 7 may not be needed for some statistics courses or may be covered in detail for other statistics courses. Please check with your instructor to determine the level at which probability will be covered. Also, unless your school has a finite math prerequisite (where the material in these three chapters would be found), you may not have been exposed to this material in your prior math courses.

PRETEST #6

Allow 20 Minutes
2 points each -- maximum score 10

6.1   Define the term *sample space*.

QUESTIONS 6.2 - 6.4 REFER TO THE FOLLOWING SITUATION:

A penny, a nickel, and a dime will be tossed simultaneously and the face of each resting coin will be recorded. Assume each coin cannot rest on its edge.

6.2   How many outcomes are possible for this experiment?

6.3   Suppose E is defined as follows: E = exactly one head is observed.
Use set notation to list the outcomes that make up event E.

6.4   Find the probability of event E.

6.5   A student who is unprepared for class is confronted with a 5-question true-false quiz as class begins. If he arbitrarily selects his answer to each question, what is the probability that he answers every question correctly?

We hear words like *chance*, *odds*, or *probability* in everyday conversation: What are my *chances* of getting that new job? What are my *odds* of winning the lottery? For today's weather, what is the *probability* of rain? In each question there was an implied condition of <u>uncertainty</u>: I don't know if I will get the new job. I'm not sure if I will win the lottery. It may or may not rain today. Whenever conditions of uncertainty appear we use probability to give us a reference scale for judging the degree of uncertainty in the situation.

This reference scale of probability can be expressed as a decimal fraction (a number between 0 and 1 -- the *chance* of getting that new job is .70), or as a percent (a number between 0% and 100% -- the probability of rain is 20%). Our job in this chapter is to understand how the framework of probability is used to produce a number on the reference scale of probability.

6.1 Vocabulary

We have already defined probability as a number on a reference scale of 0-to-1 or 0-to-100%. Before attaching meaning to the number we must understand the framework that produced the number. Two important terms in this framework are experiment and outcome, defined next:

DEFINITION - <u>Probability Framework</u>

Experiment: A process that yields potentially different outcomes.

Outcome: The simplest result that is observable but not 100% predictable.

Here are some examples of these terms:

Experiment: Toss a coin and observe the face of the resting coin.
Outcome: Head
Outcome: Tail

Experiment: Select a card from a deck of 52 playing cards and observe the suit of the selected card.

Outcome: Clubs ♣
Outcome: Diamonds ♦
Outcome: Hearts ♥
Outcome: Spades ♠

In both examples, the experiment was an activity that produced several possible outcomes. Two comments are in order about the definition of *outcome*: First, an outcome must be the <u>simplest</u> result that is observable. Note in the second example that had we defined outcomes as "a red card" or "a black card" that these would <u>not</u> have been the simplest results. Second, it should be clear that we can potentially observe a head or tail in the first example, but that we cannot predict with 100% success <u>which one</u> of these two outcomes will occur before we conduct the experiment.

The term *outcome* also may be referred to as a *simple event*, or as a *sample point* in other texts. The collection of all possible outcomes (or simple events or sample points) for an experiment is called the *sample space*. Oftentimes we can use the counting rules from Chapter 5 to help us determine the total number of sample points in the sample space. The summary example in this section will provide an illustration of this. The following definition summarizes the synonyms for some of these probability terms.

DEFINITION - <u>Probability Language Synonyms</u>

chance = probability = likelihood
outcome = simple event = sample point
sample space = universe

Set notation sometimes is used to represent the sample space and outcomes. For example, in our first experiment involving tossing a coin, we might label the outcomes as O_1 = head, and O_2 = tail, if we refer to the simplest result as an outcome, or S_1 and S_2, if we use the term sample point. The sample space, represented by the letter S, looks like the following in set notation:

$$S = \{O_1, O_2\} \quad \text{or} \quad S = \{S_1, S_2\}$$

Alternatively, your textbook may choose to represent this sample space with a figure called a *Venn diagram*, which is a graphical way of showing a sample space using a box and sample points as dots within the box as indicated below.

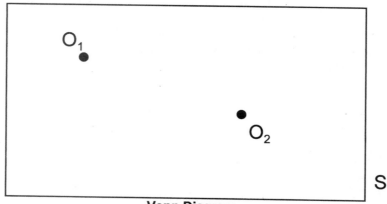

Venn Diagram

Basic Probability

An *event* (or *compound event*) is a collection of outcomes (or simple events or sample points) within a sample space. Again, set notation or Venn diagrams are often used to represent events. In our second example of selecting a card from a deck of cards, the outcomes were O_1 = clubs, O_2 = diamonds, O_3 = hearts, and O_4 = spades. If we define an event E to be a red card, then event E can be represented as follows:

$E = \{O_2, O_3\}$

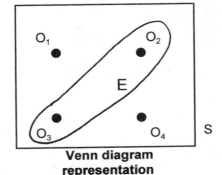

Set notation representation

Venn diagram representation

SUMMARY EXAMPLE
Listing Sample Points

A study of families with 3 children was planned in which the gender of the children would be recorded in descending birth order. Write out the sample space S as a collection of sample points. Identify the sample points that make up event E = exactly 2 of the 3 children are boys. (Assume boys and girls are equally likely.)

Do this first:	Use the product rule to count the total number of sample points	$2 \cdot 2 \cdot 2 = 8$ 1st slot = oldest child 2nd slot = middle child 3rd slot = youngest child Each slot has 2 possibilities: B = boy G = girl
Do this second:	Write out the 8 sample points	O_1 = BBB O_5 = BGG O_2 = BBG O_6 = GBG O_3 = BGB O_7 = GGB O_4 = GBB O_8 = GGG
Do this third:	Write out S as a collection of sample points	$S = \{O_1, O_2, O_3, O_4, O_5,$ $O_6, O_7, O_8\}$
Do this fourth:	Write out event E as the collection of sample points that describe E	$E = \{O_2, O_3, O_4\}$

6.2 Sample Point Approach to Probability

One approach to assigning a probability is called the <u>sample point approach to probability</u> in which we count the number of outcomes (or sample points) that make up the event and divide by the number of outcomes (or sample points) that comprise the entire sample space. The symbol for the probability of event E is P(E) and the formula is given below. Note: This formula assumes that all sample points are equally likely. All of the examples, exercises, and test questions in this chapter that ask you to compute the probability of an event are based on this equally likely assumption. In Chapter 7 we will consider the case of non-equally likely outcomes.

〰〰〰〰〰〰〰〰〰〰〰〰〰〰〰〰〰〰〰

«*FORMULA*»
PROBABILITY OF EVENT E
(SAMPLE POINT APPROACH)

$$P(E) = \frac{number\ of\ sample\ points\ in\ \ E}{number\ of\ sample\ points\ in\ \ S}$$

〰〰〰〰〰〰〰〰〰〰〰〰〰〰〰〰〰〰〰

As an example of the use of this formula consider the previous example of selecting a card from a deck of cards. The number of sample points in the sample space $S = \{O_1, O_2, O_3, O_4\}$ was 4, while the number of sample points in event E = a red card = $\{O_2, O_3\}$ was 2. Thus:

$$P(E) = \frac{number\ of\ sample\ points\ in\ \ E}{number\ of\ sample\ points\ in\ \ S}$$

$$= \frac{2}{4} = .5$$

The following summary example demonstrates the use of this formula in finding the probability of an event once the sample space has been laid out.

SUMMARY EXAMPLE
Sample Point Approach
to Probability

 A study of families with 3 children was planned in which the gender of the children would be recorded in descending birth order. Find the probability of these events:

E_1 = exactly 2 of the 3 children are boys
E_2 = all children are of the same gender
(Assume boys and girls are equally likely.)

Do this first:	Write out the sample space S (refer to previous Summary Example for the definitions of O_1 through O_8).	$S = \{O_1, O_2, O_3, O_4,$ $O_5, O_6, O_7, O_8\}$
Do this second:	Write out the collection of sample points that comprise events E_1 and E_2.	$E_1 = \{O_2, O_3, O_4\}$ $E_2 = \{O_1, O_8\}$
Do this third:	Count the number of sample points in events E_1 and E_2, and divide by the number of sample points in S.	$P(E_1) = \dfrac{3}{8} = .375$ ⬆ **Answers** ⬇ $P(E_2) = \dfrac{2}{8} = .250$

Sometimes the number of sample points in the sample space is too large to list. The same problem -- too many sample points to write out -- can occur for event E. For these situations we may have to rely on our knowledge of the counting rules from Chapter 5 to accurately count the sample points to put in the formula for the probability of event E.

For example, suppose we wanted to find the probability that in selecting 5 cards from a deck of playing cards that we would select all hearts. The experiment consists of drawing 5 cards from a deck and noting the suit of each card. Because the order in which we draw the cards is not important, the total number of sample points is:

$$_{52}C_5 = 2,598,960$$

It would take us a <u>long</u> time to write out all these sample points! Rather than trying to do this, we have to rely on our combinations rule (and our arithmetic) to give us the correct number of sample points. The same is true for event E. The number of sample points in event E is the number of different 5-card groups of hearts that can be selected from the 13 cards that are hearts:

$$_{13}C_5 = 1287$$

Finally, the probability of event E is:

$$P(E) = \frac{1287}{2,598,960} = .0004952$$

Only about 5 times in every 10,000 repetitions of drawing five cards from a deck will the result be a handful of hearts.

Chapter 6 Exercises

6.1   List all the outcomes (sample points) for the following experiment:

Experiment: Throw a pair of dice -- a white one and a red one -- and record the number of dots face-up on each die.

6.2   Each trading day the Dow Jones Industrial Average closes either higher than (Up), lower than (Down), or the same as (Same) the average from the previous day. List all the outcomes (sample points) for the following experiment:

Experiment: Record the sequential changes -- up, down, same -- in the Dow Jones Industrial Average for two consecutive days.

6.3   In a small city there are 5 candidates -- Adams, Becker, Chou, Davis, and Espinosa -- for 3 seats on the City Council. List all the outcomes (sample points) that could result when the election is held.

6.4   A bowl of marbles contains equal numbers of red, blue and orange marbles. After thoroughly mixing up the marbles in the bowl, you draw out 2 marbles, one at a time, and record the color(s). Create a Venn diagram to show the sample space for this experiment and identify all sample points.

continued on the next page ...

Basic Probability

Chapter 6 Exercises (continued)

6.5 Refer to Exercise 6.1. Identify the outcomes in the following events and then find the probability of each event

    (a) E_1 = the red die shows a 4

    (b) E_2 = the white die shows a 2 and the red die shows an odd number

    (c) E_3 = the sum of the dice is 8

6.6 Refer to Exercise 6.3. Identify the outcomes in the following events and then find the probability of each event. Assume all candidates are equally likely to be elected and that Chou is the only female candidate.

    (a) E_1 = all men are elected

    (b) E_2 = Chou is elected

    (c) E_3 = Becker is not elected

    (d) E_4 = Adams and Espinosa are elected

6.7 Refer to Exercise 6.4. Identify the outcomes in the following events and then find the probability of each event. Assume the number of marbles of each color is very large.

    (a) E_1 = both marbles are orange

    (b) E_2 = neither marble is blue

    (c) E_3 = the two marbles are not the same color

6.8 Find the probability that selecting 5 cards from a deck of cards will produce all red cards.

6.9 An auto license tag consists of 3 letters followed by 3 digits. Find the probability that a license plate begins with the letter A and ends with the digit 9.

6.10 A new car dealership employs 10 salespeople; half are men and half are women. The owner decides to select 3 of the salespeople to test drive a brand new model for a two-week period. If the owner bases her choices on chance, what is the probability that 3 women are chosen?

POSTTEST #6

Allow 20 Minutes
2 points each -- maximum score 10

6.1 Define the term *sample point*.

QUESTIONS 6.2 - 6.4 REFER TO THE FOLLOWING SITUATION:

General Motors Corporation has six major automotive divisions -- Buick, Cadillac, Chevrolet, Oldsmobile, Pontiac, and Saturn. A bonus program is established to reward all employees within the top 2 divisions that experience the lowest complaint ratio (number of complaints per 100 cars) on new cars sold. The planned per-person bonus for the number 1 ranked division is larger than the per-person bonus for the number 2 ranked division. The Cadillac and Saturn divisions are the only two divisions that build new cars in fully automated plants.

6.2 How many different outcome pairs are possible from the six divisions?

6.3 Suppose E is defined as follows:

E = Neither of the top 2 ranking divisions has a fully automated plant.
Use set notation to list the outcomes that make up event E.

6.4 Find the probability of event E.

6.5 The FCC assigns each airport in the United States a 3-letter code. What is the probability of correctly guessing the code for a given airport if one knows only the first letter of the code for sure?

CHAPTER SEVEN
Intermediate Probability

NOTE TO STUDENTS

The probability material in Chapters 5, 6, and 7 may not be needed for some statistics courses or may be covered in detail for other statistics courses. Please check with your instructor to determine the level at which probability will be covered. Also, unless your school has a finite math prerequisite (where the material in these three chapters would be found), you may not have been exposed to this material in your prior math courses.

PRETEST #7

Allow 30 Minutes
1 point each part (Questions 1 & 2)
2 points each part (Question 3) -- maximum score 10

7.1   A sample space and some events are defined as the following collection of equally likely sample points:

$$S = \{O_1, O_2, O_3, O_4, O_5, O_6, O_7\}$$
$$E_1 = \{O_2, O_3, O_7\}$$
$$E_2 = \{O_1, O_4, O_5, O_6, O_7\}$$
$$E_3 = \{O_1, O_4, O_6\}$$

Write out the set of sample points that make up the following events:

(a) $E_1 \cap E_2$ **(b)** $E_2 \cup E_3$ **(c)** E_2^c **(d)** $E_1 \cap E_3$

7.2   Refer to Question 7.1. Find the probability of each event described in parts (a) through (d).

7.3   The probability of event A is .30, and the probability of event B is .60. Events A and C do not intersect. The probability of event C, given that event B has occurred is .50. The probability of events A and B is .20. Find the probability of...

(a) the complement of event B.
(b) events A or B.
(c) events B and C.
(d) event B, given that event A has occurred.
(e) Are events A and C mutually exclusive events? Why?
(f) Are events A and B independent events? Why?

By listing all the sample points in a sample space we easily can identify those sample points that make up an event, as we saw in the previous chapter. This idea of defining an event as a collection of sample points can be extended to new events that are related to old events or are formed by combining old events. There are three main ways to form new events: union, intersection, and complement.

7.1 Relating and Combining Events

Let us understand the framework from which we are working. We have a sample space S, which is composed of many sample points (or outcomes) that are numbered from 1 to n. Two (or more) events -- E_1 and E_2 -- are identified as containing some, but not all, of the sample points in S. We would like to create new events by combining E_1 and E_2 or by relating an event to one or the other or both events

One way to combine the events into a new event is to merge them together; this is called forming the *union* of the two events. The new event is not given a new letter (such as E_3 or F), but is represented as the symbol "$E_1 \cup E_2$," where the capital letter U denotes union.

DEFINITION - Union of Two Events

The union of events E_1 and E_2 is represented by the symbol $E_1 \cup E_2$ and merges all the sample points in event E_1 with those in E_2.

In listing the merged sample points in $E_1 \cup E_2$ we do <u>not</u> repeat any sample points that happen to occur in both events. For instance, if $E_1 = \{O_1, O_3, O_5\}$ and $E_2 = \{O_1, O_2\}$, then $E_1 \cup E_2 = \{O_1, O_2, O_3, O_5\}$. We do not need to list the outcome O_1 twice; once is enough.

A second way of creating a new event is to filter them to find sample points that they share in common; this is called forming the *intersection* of the two events. The new event is represented as the symbol "$E_1 \cap E_2$," where the upside down capital letter U denotes intersection.

DEFINITION - Intersection of Two Events

The <u>intersection</u> of events E_1 and E_2 is represented by the symbol $E_1 \cap E_2$ and contains all the sample points that events E_1 and E_2 share in common.

For instance, if $E_1 = \{O_1, O_3, O_5\}$ and $E_2 = \{O_1, O_2\}$, then $E_1 \cap E_2 = \{O_1\}$ because O_1 is the only sample point common to both E_1 and E_2.

It may happen that events E_1 and E_2 have no outcomes in common -- none of the sample points in either event matches. When no sample point is included in both events simultaneously, we say that the intersection is empty or that the events are mutually exclusive of one another. The symbol from set notation to indicate an empty intersection is a zero with a slash through it: $\emptyset$ (the empty set).

DEFINITION - <u>Mutually Exclusive Events</u>

When events E_1 and E_2 share no sample points in common we call them <u>mutually exclusive</u> events. Another way to characterize mutually exclusive events is to say their intersection is empty: $E_1 \cap E_2 = \emptyset$.

A final way that we will consider to create a new event is a variation of the "half-empty glass of water" idea. Perhaps you are familiar with the two ways of viewing a glass that is filled halfway up with water. One way describes the glass as "half-full," while an opposite point of view describes the same glass as "half-empty." Note that the phrases "half-full" and "half-empty" are opposites of one another, yet together they define the capacity of the glass. Or, we could define one part of the glass in terms of the other: The half-empty part is the part of the glass that does not contain water.

A new event can be created from an old event E in a similar manner by defining the new event as that part of the sample space that does not contain the old event; this is called forming the *complement* of the event. The new event is represented by the symbol E^c. (Other texts might use the symbols E' or $\overline{E}$ in place of E^c.)

DEFINITION - <u>Complement of an Event</u>

If event E contains some of the sample points in S, then the complement of event E, represented by the symbol E^c, contains the remaining sample points in S. Another way to characterize E^c is to say that it contains all sample points in S that are not in E.

Venn diagrams show the creation of these new events through pictures. Next are four panels that illustrate each of the definitions that we just presented. The shaded area, where appropriate, is meant to represent the newly created event.

Intermediate Probability

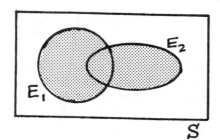

UNION

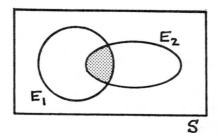

INTERSECTION

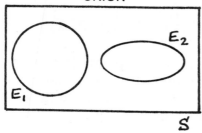

MUTUALLY EXCLUSIVE

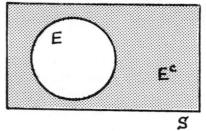

COMPLEMENT

SUMMARY EXAMPLE
Relating and Combining Events

A sample space S contains 10 sample points: $S = \{O_1, O_2, O_3, O_4, O_5, O_6, O_7, O_8, O_9, O_{10}\}$
Events E_1, E_2, and E_3 are defined as the following collection of sample points:

$E_1 = \{O_2, O_6, O_7, O_8, O_9, O_{10}\}$ $E_2 = \{O_1, O_3, O_5\}$ $E_3 = \{O_2, O_4, O_5, O_6\}$

Write out the set of sample points that make up the following newly created events:

(1) $E_2 \cup E_3$ (2) $E_1 \cap E_3$ (3) E_1^c

(4) Are events E_1 and E_2 mutually exclusive events? Why?

Part (1)	Merge sample points in E_2 and E_3	$E_2 \cup E_3 = \{O_1, O_2, O_3, O_4, O_5, O_6\}$
Part (2)	Match common sample points in E_1 and E_3	$E_1 \cap E_3 = \{O_2, O_6\}$
Part (3)	E_1 accounts for 6 sample points in S. Identify the remaining, unaccounted for sample points in S.	$E_1^c = \{O_1, O_3, O_4, O_5\}$
Part (4)	Answer the question, noting that E_1 and E_2 have no sample points in common.	Yes, because $E_1 \cap E_2 = \emptyset$

As we mentioned in Chapter 6, calculating probabilities by counting the sample points in the sample space and dividing by the total number of sample points is based on the assumption of equally likely outcomes. However, <u>not</u> all sample spaces are composed of equally likely outcomes. For example, suppose we had a bag of peanut M&M candies and we were interested in selecting one peanut candy from the bag and identifying its color (before we eat it!). The sample points might be

$$
\begin{aligned}
O_1 &= \text{brown} \\
O_2 &= \text{yellow} \\
O_3 &= \text{orange} \\
O_4 &= \text{green} \\
O_5 &= \text{red}
\end{aligned}
$$

so that the sample space is S = {O_1, O_2, O_3, O_4, O_5}. You may or may not be aware of the mix of colors of all peanut M&Ms that are made: 30% brown, 20% yellow, 10% orange, 20% green, and 20% red. Thus, the five sample points are not equally likely. In this case we would assign the following probabilities to the sample points:

$$
\begin{aligned}
P(O_1) &= .30 \\
P(O_2) &= .20 \\
P(O_3) &= .10 \\
P(O_4) &= .20 \\
P(O_5) &= .20
\end{aligned}
$$

If event E were defined as the collection of sample points O_2, O_3, and O_4:

$$ E = \{O_2, O_3, O_4\} $$

we would find the probability of E by adding the probabilities of the individual sample points (rather than counting the sample points and dividing by 5):

$$ P(E) = P(O_2) + P(O_3) + P(O_4) = .10 + .20 + .20 = .50 $$

7.2 Rules of Probability

In the previous section we learned how to create new events by relating or combining old events. With a list of all the sample points in the sample space it was not difficult to write out the union, intersection, or complement of events as collections of sample points. Now that we can <u>see</u> how the new events are formed either through Venn diagrams or by specifying the outcomes in the events, we wish to extend our knowledge to situations where the sample points are not specifically spelled out. In particular, we want to review the probability rules that apply to our new events so that we can find probabilities of unions, intersections, and complements even when the sample points are not listed.

The first probability rule is the *additive rule* and it applies to the *union* of events. (Synonyms for the additive rule are the addition rule or the union rule.) Usually the rule is stated in terms of events A and B rather than events E_1 and E_2.

RULES
ADDITIVE RULE OF PROBABILITY

$$P(A \cup B) = P(A) + P(B) - P(A \cap B)$$

Although this rule can be applied to all sample spaces, we may not find it particularly useful or efficient if it's possible to count sample points instead. For example, in the Summary Example of the previous section we identified the union of events E_2 and E_3 as a set of 6 outcomes from the sample space. By merely counting (and not using the additive rule) we can state that $P(E_2 \cup E_3) = {}^6/_{10} = .6$. Whenever possible, we recommend counting the sample points over using the additive rule (as well as the subsequent rules to follow) to find the probability of $A \cup B$.

A situation for which we cannot count sample points occurs when the probabilities of the individual events are stated without a complete delineation of the sample points in the sample space. For example, suppose we were told that $P(A) = .70$, $P(B) = .50$ and $P(A \cap B) = .25$. Then to find $P(A \cup B)$, we apply the additive rule:

$$\begin{aligned} P(A \cup B) &= P(A) + P(B) - P(A \cap B) \\ &= .70 + .50 - .25 = .95 \end{aligned}$$

In this problem there were no sample points given; instead, the probabilities of the separate events plus the intersecting events were given.

A second probability rule is a special case of the first rule and a consequence of events being mutually exclusive. Recall that mutually exclusive events do not intersect; thus, the probability of their intersection must be 0. That is, if events A and B are mutually exclusive, then $P(A \cap B) = 0$. This means that in the additive rule the last term in the equation would disappear.

RULES
ADDITIVE RULE OF PROBABILITY
FOR MUTUALLY EXCLUSIVE EVENTS

$$P(A \cup B) = P(A) + P(B)$$

The third probability rule is the *complementary rule* and applies to the complement of an event. This rule formalizes the idea that the half-full and half-empty parts of the glass combine to define the total capacity of the glass.

RULES

COMPLEMENTARY RULE OF PROBABILITY

$$P(A) + P(A^c) = 1$$
Alternative forms of the rule are: $P(A) = 1 - P(A^c)$ and $P(A^c) = 1 - P(A)$

The fourth probability rule applies to the intersection of events, but we will wait to introduce it in the next section after we have discussed the concept of conditional probability.

SUMMARY EXAMPLE
Rules of Probability

In a five-county region surrounding and encompassing a medium-sized city there are three different companies -- A, B, and C -- that provide cellular telephone service. Company A covers 80% of the five-county region. Company B covers 60% of the region, while Company C's coverage is only 25%. In 50% of the five-county region Company A's coverage overlaps with Company B's coverage. Companies B and C have no areas of the region where they compete head-to-head. If coverage percents can be viewed as probabilities, find the probability of these events: (1) $P(A \cup B)$ (2) $P(B \cup C)$ (3) $P(A^c)$

Preliminary activity:	Write down the given information	$P(A) = .80$ $P(B) = .60$ $P(C) = .25$ $P(A \cap B) = .50$ $P(B \cap C) = 0$
Part (1)	Apply the additive rule	$P(A \cup B) = P(A) + P(B) - P(A \cap B)$ $= .80 + .60 - .50$ $= \mathbf{.90} \leftarrow \mathbf{Answer}$
Part (2)	Because $P(B \cap C) = 0$, apply the additive rule for mutually exclusive events	$P(B \cup C) = P(B) + P(C)$ $= .60 + .25$ $= \mathbf{.85} \leftarrow \mathbf{Answer}$
Part (3)	Apply the complementary rule	$P(A^c) = 1 - P(A)$ $= 1 - .80$ $= \mathbf{.20} \leftarrow \mathbf{Answer}$

7.3 Conditional Probability

Up to this point in our study of probability we have been working with *unconditional* probability statements only. Now we wish to consider what happens when limiting factors or conditions are present. You might think of these factors or conditions as additional information. Sometimes this added knowledge affects the uncertainty of the event; sometimes it doesn't.

For example, an analysis of mortgage loans in the banking industry revealed possible evidence of discrimination against African-Americans. According to banking regulations, the approval of a mortgage loan application is supposed to be a nondiscriminatory process where applicants are approved provided they fall within certain financial ratio guidelines. However, an examination of applications filed by African-Americans revealed a smaller proportion of loan approvals than the proportion for the general public. (What remained to be analyzed was whether loans were denied despite favorable financial ratios.)

In the language of conditional probability for this example, we might define events A and B as follows:

A: mortgage loan application is approved
B: applicant is African-American

What the banking analysis revealed was that the uncertainty of event A was influenced by event B. To represent this situation in probability symbols, we introduce the notion of a *conditional probability* statement as $P(A|B)$, which is read as "the conditional probability that event A occurs, given that event B has occurred." The (unconditional) probability of event A is just $P(A)$. Thus the banking analysis found that $P(A|B)$ was a smaller number than $P(A)$.

As was true with the probability rules in the previous section, it is easy to find a conditional probability by counting sample points, provided a listing of all the sample points in the sample space is given. For example, suppose:

$$S = \{O_1, O_2, O_3, O_4, O_5\}$$
$$A = \{O_1, O_3, O_5\}$$
$$B = \{O_1, O_2, O_4, O_5\}$$

The (unconditional) probability of event A is $P(A) = {}^3/_5 = .6$. To find $P(A|B)$ we must remember that the conditional information is that "event B has occurred." Note the past tense of that phrase. If event B has occurred, then the only possible outcomes are $O_1, O_2, O_4,$ and O_5; outcome O_3, which is <u>not</u> a part of event B, is eliminated. In other words, think of event B as the reduced sample space containing 4 outcomes. How many of these 4 outcomes are part of event A? With the elimination of outcome O_3, there are only 2 remaining outcomes in event A -- O_1 and O_5 -- that are included in the reduced sample space of event B. Thus, $P(A|B) = {}^2/_4 = .5$.

If sample points are not explicitly listed, then the following formula shows how to calculate a conditional probability, assuming $P(B) \neq 0$:

«*FORMULA*»
CONDITIONAL PROBABILITY

$$P(A \mid B) = \frac{P(A \cap B)}{P(B)}$$

This formula requires us to know the probability of the intersection of events A and B plus the (unconditional) probability of event B in order to find $P(A \mid B)$.

Rearranging this formula produces the fourth (and last) rule of probability known as the *multiplicative rule*, which applies to the intersection of events.

RULES
MULTIPLICATIVE RULE OF PROBABILITY

$P(A \cap B) = P(A \mid B) \, P(B)$
Reversing the roles of A and B in the above equation
yields the alternative form of the rule $P(A \cap B) = P(B \mid A) \, P(A)$

Recall we said that sometimes the added knowledge in a conditional probability affects the uncertainty of the event and sometimes it doesn't. This means in some problems we find that $P(A)$ and $P(A \mid B)$ are different numbers, and in other problems we find that $P(A)$ and $P(A \mid B)$ are the same number. These different conditions give rise to a couple of descriptive terms:

DEFINITION - <u>Independent and Dependent Events</u>

When $P(A)$ and $P(A|B)$ are the same number, we call events A and B <u>independent</u> events.

When $P(A)$ and $P(A|B)$ are different numbers, we call events A and B <u>dependent</u> events.

The roles of events A and B are interchangeable in the previous definition, so that we could compare $P(B)$ to $P(B|A)$. If $P(B)$ and $P(B|A)$ are the same number, events A and B are independent. If $P(B)$ and $P(B|A)$ are different numbers, events A and B are dependent.

Finally, in light of the above definition of independent events the fourth rule of probability can be edited slightly (as long as events A and B are independent events, not dependent events).

RULES
MULTIPLICATIVE RULE OF
PROBABILITY FOR INDEPENDENT EVENTS

$$P(A \cap B) = P(A)\,P(B)$$

In closing this chapter we would like to offer a hint for working probability problems that involve the use of the probability rules. Obviously, if the problem lists sample points, we encourage you to find probabilities by counting the sample points (instead of applying the rules). If there are no sample points and the problem is stated in symbolic form, such as $P(A \cup B)$ or $P(C \cap D)$, then apply the probability rules directly. But if the problem has no sample points and no symbols -- just words -- then look for <u>key words</u> to decide which probability rule to use. The following word associations may help:

KEY WORD	OPERATION	APPROPRIATE PROBABILITY RULE TO USE	EXAMPLE
Or	Union	Additive	... find the probability of this <u>or</u> that ...
And	Intersection	Multiplicative	... find the probability of both this <u>and</u> that ...
Not	Complement	Complementary	... find the probability of <u>not</u> this ...

SUMMARY EXAMPLE
Conditional Probability

A survey of women who have pre-school age children revealed the following breakdown (WOTH = working outside the home):

    ✔ married and WOTH, 60%   ✔ married and not WOTH, 15%
    ✔ not married and WOTH, 20%   ✔ not married and not WOTH, 5%

Define events A and B as: **A:** married   **B:** WOTH and assume P(A) = .75 and P(B) = .80

    Find the following probabilities:
    (1) $P(A \cap B)$
    (2) $P(A|B)$
    (3)   Are events A and B independent events? Why?
    (4)   What probability symbol describes the 15% figure in the second line of the breakdown?

Part (1)	Associate intersection with the key word "and"	$P(A \cap B) = $ **.60** ← **Answer**
Part (2)	Use multiplicative rule and answer to first part	$P(A|B) = \dfrac{P(A \cap B)}{P(B)}$ $= \dfrac{.60}{.80} = $ **.75** ← **Answer**
Part (3)	Compare values for P(A) and P(A|B) from the second part	Yes, because P(A) and P(A|B) are the same number (.75)
Part (4)	Interpret the key word "and" as *intersection* and the key word "not" as *complement*	$P(A \cap B^c)$ ← **Answer**

Chapter 7 Exercises

7.1   A sample space and some events are defined as the following collection of equally likely sample points:

$$S = \{O_1, O_2, O_3, O_4, O_5, O_6, O_7, O_8\}$$
$$E_1 = \{O_1, O_8\}$$
$$E_2 = \{O_2, O_5, O_6, O_8\}$$
$$E_3 = \{O_3, O_4, O_5, O_6, O_7, O_8\}$$

Write out the set of sample points that make up the following events:

(a) E_3^c

(b) $E_1 \cup E_2$

(c) $E_2 \cap E_3$

(d) $E_3^c \cap E_2$

(e) $E_1 \cap E_2 \cap E_3$

(f) $E_2^c \cup E_3$

(g)   Are events E_1 and E_2 mutually exclusive events? Why?

7.2   A sample space and some events are defined as the following collection of equally likely sample points:

$$S = \{O_1, O_2, O_3, O_4, O_5, O_6, O_7, O_8, O_9, O_{10}, O_{11}, O_{12}\}$$
$$E_1 = \{O_2, O_4, O_8, O_{11}\}$$
$$E_2 = \{O_1, O_3, O_5, O_6, O_7\}$$
$$E_3 = \{O_1, O_4, O_7, O_8, O_{12}\}$$
$$E_4 = \{O_9, O_{10}\}$$

Write out the set of sample points that make up the following events:

(a) $E_1 \cup E_4$
(b) $E_2 \cap E_3$
(c) $E_1 \cap E_3$
(d) E_2^c
(e) $E_2 \cup E_4$
(f) $E_1^c \cap E_3^c$
(g)   Are events E_1 and E_2 mutually exclusive events? Why?
(h)   Are events E_3 and E_4 mutually exclusive events? Why?

7.3   Refer to Exercise 7.1. Find the probability of the event described in parts **(a)** through **(f)**.

7.4   Refer to Exercise 7.2. Find the probability of the event described in parts **(a)** through **(f)**.

continued on the next page ...

Chapter 7 Exercises (continued)

7.5 Refer to Exercise 7.1. Find these probabilities:

(a) $P(E_1|E_3)$
(b) $P(E_3|E_2)$
(c) Are events E_1 and E_3 independent events? Why?

7.6 Refer to Exercise 7.2. Find these probabilities:

(a) $P(E_2|E_4)$
(b) $P(E_3|E_1)$
(c) $P(E_2|E_3)$
(d) Are events E_1 and E_3 independent events? Why?
(e) Are events E_2 and E_4 independent events? Why?

7.7 The shaded area in each of the following Venn diagrams depicts a new event that is a combination -- union, intersection, complement -- of the events A, B, and C. Express the new event in symbols. Note: all letters (A and B in parts a, b, and d, and A, B, and C in part c) should be used in your answer.

(a)

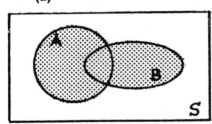

(b)

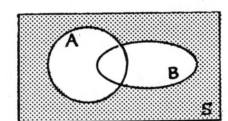

(c)

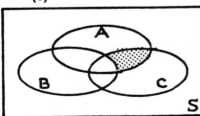

(d)

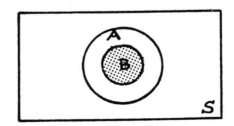

continued on the next page ...

Chapter 7 Exercises (continued)

7.8   Find the indicated probabilities using the appropriate probability rule and the following information:

$$P(A) = .10 \qquad\qquad P(A \cap B) = 0$$
$$P(B) = .20 \qquad\qquad P(A \cap C) = .03$$
$$P(C) = .60 \qquad\qquad P(B|C) = .25$$

(a) $P(A \cup B)$
(b) $P(A \cup C)$
(c) $P(B \cap C)$
(d) $P(B|A)$
(e) $P(A|C)$
(f) $P(C|A)$
(g)   Are events A and B independent events? Why?
(h)   Are events A and C independent events? Why?

7.9   Find the indicated probabilities using the appropriate probability rule and the following information:

$$P(A) = .80 \qquad P(C|A) = .25 \qquad P(A \cap D) = .30$$
$$P(C) = .25 \qquad P(D|B) = .25 \qquad P(B \cap C) = .05$$
$$P(D) = .35 \qquad P(B \cup C) = .40$$

(a) $P(A \cup D)$
(b) $P(B|C)$
(c) $P(A \cap C)$
(d) $P(D|A)$
(e) $P(B)$
(f) $P(B \cap D)$
(g)   Are events A and C independent events? Why?
(h)   Are events A and D independent events? Why?
(I)   Are events B and C independent events? Why?

continued on the next page ...

Intermediate Probability

Chapter 7 Exercises (continued)

7.10 A study of a graduating class of college seniors revealed the following information about their gender and dexterity:

> 85% were right-handed only
> 5% were ambidextrous
> 10% were male and left-handed only
> 50% were female
> 45% were female and right-handed only

All students in the study were either right-handed only or left-handed only or ambidextrous (both right- and left-handed). With the following definitions of events, find the indicated probabilities:

> M = male   R = right-handed only   A = ambidextrous
> F = female   L = left-handed only

(Hint: Create a table with three columns labeled R, L, and A, and two rows labeled M and F.)

(a) P(M)
(b) P(L)
(c) P(F ∩ L)
(d) P(F|A)
(e) P(L|M)
(f) P(R|F)

7.11 A sample space and some events are defined as the following collection of <u>non</u>-equally likely sample points:

$$S = \{O_1, O_2, O_3, O_4, O_5, O_6\}$$
$$E_1 = \{O_1, O_2, O_5\}$$
$$E_2 = \{O_3, O_5, O_6\}$$

Each of the odd-numbered sample points has probability equal to .10, while $P(O_6) = .30$, and the remaining sample points each have probability equal to .20. Find these probabilities:

(a) $P(E_1)$
(b) $P(E_1 \cap E_2)$
(c) $P(E_1|E_2)$

continued on the next page ...

Chapter 7 Exercises (continued)

7.12 An in-home pregnancy test yielded the following results in an extensive field test (positive test reading implies pregnancy):

Actual Clinical State	Test Reading	
	Positive	Negative
Pregnant	.75	.05
Not pregnant	.02	.18

Note: the number .75 means that 75% of the women who participated in the field test were pregnant and had a positive test reading. Also, 80% of the women who participated in the study were actually pregnant (.80 = .75 + .05). Events are defined as follows:

A = pregnant   B = not pregnant
S = positive test reading   N = negative test reading

Find the indicated probabilities:

(a) P(B)
(b) P(S)
(c) P(B ∩ N)
(d) P(A|S)
(e) P(N|B)
(f) Express the following phrase in symbols: the probability that a test reading will be positive, given that the woman is pregnant.

7.13 A survey of eating habits of adult Americans was segmented into three age groups: young adults, middle-aged, and senior citizens. The young adult and middle-aged segments each accounted for 40% of the respondents. One of the survey findings showed that 90% of adult Americans eat meat, while 10% of adult Americans are vegetarians. Only 19% of those surveyed were classified as a senior and a meat eater. Given that a respondent was a vegetarian, there was a 50% chance that he/she was a young adult. It was also discovered that the events "vegetarian" and "middle-aged" were independent events. What is the probability of selecting a respondent who is...

(a) a meat eater?
(b) not a young adult?
(c) middle-aged, given that he/she is a vegetarian?
(d) a young adult and a vegetarian?
(e) a sénior or a meat eater?

(HINT: Create a table with three columns labeled Young, Middle-aged, and Senior, and two rows labeled Meat and Vegetarian.)

POSTTEST #7

Allow 30 Minutes
1 point each part (Questions 1 & 2)
2 points each part (Questions 3-6) --
maximum score 26

7.1 If we form the union of an event and its complement, how much is the probability of the new event?

7.2 A sample space and some events are defined as the following collection of equally likely sample points:

$$S = \{O_1, O_2, O_3, O_4, O_5, O_6, O_7, O_8, O_9, O_{10}\}$$
$$E_1 = \{O_1, O_2, O_3, O_4, O_{10}\}$$
$$E_2 = \{O_5, O_6, O_7\}$$
$$E_3 = \{O_1, O_3, O_5, O_7, O_9, O_{10}\}$$

Write out the set of sample points that make up the following events:

(a) E_3^c
(b) $E_2 \cup E_3$
(c) $E_1 \cap E_3$

7.3 Refer to Question 7.2. Find these probabilities:

(a) $P(E_3|E_2)$
(b) $P(E_1|E_3)$

7.4 Refer to Question 7.2.

(a) Are events E_1 and E_2 mutually exclusive events? Why?
(b) Are events E_1 and E_3 independent events? Why?

continued on the next page...

7.5   The owner of a music store recorded the primary musical preference (country, rock, or other) and gender of her customers and found the following:

Prefer country music 50%
Prefer rock music and is male customer 20%
Prefer country music and is female customer 20%
Is female customer 40%
Does not prefer rock music 65%
Prefer country music, given a male customer 50%

Find the probability that a customer...

(a)   is male or prefers rock music.
(b)   prefers country music, given that the customer is female.
(c)   is male and prefers country music.
(d)   Are the events "male customer" and "prefers country music" independent events? Why?

7.6   The average bag of M&Ms plain chocolate candies contains 30% browns, 20% yellows, 20% reds, 10% oranges, 10% greens, and 10% tans. If we select one piece of candy from a bag, let events E_1, E_2, and E_3 be defined as follows:

E_1 = {brown, red, orange, green}
E_2 = {brown, tan}
E_3 = {red, orange, yellow}

Find the probability of the following events:

(a) $P(E_2)$
(b) $P(E_1|E_3)$
(c) $P(E_1 \cup E_2)$

PRETEST #8
Allow 25 Minutes
1 point each -- maximum score 7

IN QUESTIONS 8.1 - 8.4 SOLVE THE EQUATION FOR X

8.1 $12 - 5X = -8$

8.2 $\dfrac{10X + 4}{X - 5} = 4$

8.3 $3 + \dfrac{12}{X} = 5$

8.4 $.6 = \dfrac{1.5(4)}{\sqrt{X}}$

8.5 Solve the following equation for X: $-4X + 16Y = -24$

8.6 Solve the following inequality for X: $12 - 3X > -3$

8.7 All cotton sweaters, regularly priced at $39.99, were moved to clearance and discounted 20%. What is the most number of sweaters you can buy if you have $100 to spend? (Ignore sales tax.)

An *equation* is a statement of equality relating variables and constants. We recognize variables by letters such as X or Y; constants are numbers that stand alone or are positioned beside variables as multipliers of the variables. For example, in the equation: $2X + 3 = 7$ the number 3 and 7 are *constants*, X is the *variable*, and 2 is a constant called the *coefficient of the variable*. Typically we are interested in <u>solving the equation</u> for X.

Solving an equation for X means we wish to isolate the variable X on one side or the other of the equals sign. When X is all alone on the left side of the equals sign (for example), the number on the right side of the equals sign is called the *solution value* for the equation. The solution value is said to *satisfy the equation*. Although this discussion revolves around getting a <u>number</u> for the solution value, we can also solve an equation that has several variables in it for one of the variables. In this case the solution value will necessarily involve other variables and will not lead to a single number answer.

For instance, if we solve the equation $2X + 3 = 7$ for X we get the solution value $X = 2$. But if we solve the equation $3X + 12Y = 3$ for X we get the solution value $X = -4Y + 1$.

8.1   Solution Technique

The basic premise in solving an equation is that when we perform the same operation to both sides of the equation we do not affect the solution value for the equation. For instance, if we had an equation such as:

$$X - 3 = 10$$

then by adding 3 to both sides of the equation we eliminate the -3 on the left side and reveal the solution value for the equation in the process:

$$X - 3 + 3 = 10 + 3$$
$$X = 13$$

Similarly, if we started with $2X = 10$ and divided both sides by 2, we would get the solution value $X = 5$.

Our goal in solving an equation is to repeatedly use the premise of "doing the same thing to both sides of the equals sign" in order to isolate the variable on one side. Here is an example of this process: Solve the following equation for X:

$$X + 4(1 - 2X) = 2(2 - X) + 5$$

First, perform the indicated multiplications, as required by the priority of arithmetic operations in Chapter 1:

$$X + 4 - 8X = 4 - 2X + 5$$

Second, combine the X and -8X terms on the left side, and the constants 4 and 5 on the right side of the equals sign:

$$-7X + 4 = 9 - 2X$$

Third, subtract 4 from both sides of the equation. This will eliminate the 4 on the left side:

$$-7X + 4 - 4 = 9 - 2X - 4$$
$$-7X = 5 - 2X$$

Fourth, add 2X to both sides of the equation. This will eliminate the -2X term on the right side:

$$-7X + 2X = 5 - 2X + 2X$$
$$-5X = 5$$

Finally, divide both sides of the equation by -5 to get X = -1 as the solution value:

$$\frac{-5X}{-5} = \frac{5}{-5} \quad \rightarrow \quad X = -1$$

 We <u>always</u> recommend that you check the solution value to ensure that it satisfies the original equation. To do this go back to the original equation and replace all occurrences of the variable with the solution value and then do the indicated arithmetic to verify that the left side of the original equation produces the same number as on the right side. In our example, we would replace all occurrences of the variable X with the number -1:

Left side	$\overset{?}{=}$	Right side
X + 4(1 - 2X)	$\overset{?}{=}$	2(2 - X) + 5
(-1) + 4(1 - 2(-1))	$\overset{?}{=}$	2(2 - (-1)) + 5
-1 + 4(1 + 2)	$\overset{?}{=}$	2(2 + 1) + 5
-1 + 4(3)	$\overset{?}{=}$	2(3) + 5
-1 + 12	$\overset{?}{=}$	6 + 5
11	$\checkmark$	11

Because the left side value of 11 matches the right side value of 11, then we have verified that X=-1 is indeed the solution value of the equation.

In verifying that the left side equals the right side, we encourage you <u>not</u> to move terms from one side of the equals sign to the other side and <u>not</u> to multiply or divide both sides by a constant. Instead, just work down the left side and separately work down the right side to produce the same number. Between the left side and the right side we included a symbol that shows an = sign with ? on top of it. This is to indicate that we don't know if the left side will equal the right side. In the last line we replaced the ? with a ✓ to show that the uncertainty is now removed and that the left side value does indeed match the right side value.

SUMMARY EXAMPLE
Solution Technique

Solve the following equation for X:

$$\frac{14X - 3}{5} = 2X + 1$$

Do this first:	Multiply both sides by 5; this will eliminate the 5 on the left side	$5\dfrac{(14X - 3)}{5} = 5(2X + 1)$ $14X - 3 = 10X + 5$
Do this second:	Add 3 to both sides; this will eliminate the -3 on the left side	$(14X - 3) + 3 = (10X + 5) + 3$ $14X = 10X + 8$
Do this third:	Subtract 10X from both sides; this will eliminate the 10X on the right side	$(14X) - 10X = (10X + 8) - 10X$ $4X = 8$
Do this fourth:	Divide both sides by 4; this will eliminate the 4 on the left side	$\dfrac{(4X)}{4} = \dfrac{(8)}{4}$ X = 2 ← **Answer**
Do this fifth:	Verify that X = 2 satisfies the equation. Replace all occurrences of X with the number 2 and perform the arithmetic.	$\dfrac{14X - 3}{5} \overset{?}{=} 2X + 1$ $\dfrac{14(2) - 3}{5} \overset{?}{=} 2(2) + 1$ $\dfrac{28 - 3}{5} \overset{?}{=} 4 + 1$ $\dfrac{25}{5} \overset{\checkmark}{=} 5$

Solving an Equation

In the second step in the Summary Example (on previous page), we added 3 to both sides of the equation. The mathematical slang for this procedure is to "take 3 to the other side." If this is how you learned to solve equations, then we wish to remind you that taking 3 from the left side to the right side changes the sign from -3 on the left side to +3 on the right side. Similarly in the first step, you might be more familiar with the command "crossmultiply by 5." Again, we remind you that crossmultiplying by 5 takes the 5 from the <u>bottom</u> of the left side and moves it across the equals sign to the <u>top</u> of the right side.

8.2 Solving the Equation $Z = \dfrac{X - \mu}{\sigma}$

In statistics, one equation that we might encounter is the following expression called a *Z-score*:

$$Z = \frac{X - \mu}{\sigma}$$

This Z-score equation likely will appear in your statistics text in a section or chapter dealing with the normal probability distribution. On first glance, we see some strange symbols, μ and σ, which are letters of the Greek alphabet (lower case mu and sigma, respectively), and we do not see any constants. Instead, it looks like the equation has four unknowns in it. We cannot solve such an equation for a single solution value without knowing the values of three of the four unknowns. For instance, if we knew that $X = 22$, $\mu = 19$, and $\sigma = 5$, then we can solve the equation for the variable Z and get:

$$Z = \frac{X - \mu}{\sigma} = \frac{22 - 19}{5} = \frac{3}{5} = .6$$

In a sense, the equation:

$$Z = \frac{X - \mu}{\sigma}$$

is already solved for Z; all that remains is to drop in the values of X, μ, and σ, as indicated above. However, this same equation can be rearranged and solved in a similar manner for the other symbols. For example, let's solve the equation for X:

$$Z = \frac{X - \mu}{\sigma}$$

First, multiply both sides by σ; this will eliminate the σ on the right side:

$$(Z)\sigma = (\frac{X - \mu}{\sigma})\sigma$$
$$Z\sigma = X - \mu$$

Second, add μ to both sides; this will eliminate the μ on the right side:

$$(Z\sigma) + \mu = (X - \mu) + \mu$$
$$Z\sigma + \mu = X$$

Third, reverse the left side and the right side: $\quad X = Z\sigma + \mu$

This is the solution value for the original equation and expresses X in terms of the other variables. To get a number answer for X, we would have to be given number values for Z, σ, and μ. For example, if $Z = 1.2$, $\mu = 24$, and $\sigma = 2$, what is the solution value for X?

$$X = Z\sigma + \mu = (1.2)(2) + 24$$
$$= 2.4 + 24 = 26.4$$

The solution value is $X = 26.4$, when $Z = 1.2$, $\mu = 24$, and $\sigma = 2$.

8.3 Solving the Equation $E = Z \dfrac{\sigma}{\sqrt{n}}$

Another equation that you will see in your statistics textbook when you are studying confidence interval estimation for μ is:

$$E = Z\frac{\sigma}{\sqrt{n}}$$

Again, we are faced with one equation having four unknowns and no constants. Since the equation is already solved for E, in general terms anyway, we will assume that, given values for Z, σ, and n, we can generate the solution value for E.

Of more importance to us will be solving the equation for n. First, multiply both sides by $\sqrt{n}$; this will eliminate the $\sqrt{n}$ from the denominator of the right side:

$$(E)\sqrt{n} = \left(Z\frac{\sigma}{\sqrt{n}}\right)\sqrt{n}$$
$$E\sqrt{n} = Z\sigma$$

Second, divide both sides by E; this will eliminate the E on the left side and isolate $\sqrt{n}$ on one side:

$$\frac{(E\sqrt{n})}{E} = \frac{(Z\sigma)}{E}$$
$$\sqrt{n} = \frac{Z\sigma}{E}$$

Solving an Equation

Our objective was to solve for n, not $\sqrt{n}$. Recall from Chapter 4 that we can express $\sqrt{n}$ as $n^{\frac{1}{2}}$ and that squaring $n^{\frac{1}{2}}$ will produce n: $(n^{\frac{1}{2}})^2 = n^1 = n$. Finally, in the above equation, square both sides; this will turn $\sqrt{n}$ into n:

$$(\sqrt{n})^2 = \left(\frac{Z\sigma}{E}\right)^2$$

$$n = \left(\frac{Z\sigma}{E}\right)^2$$

In statistics, this equation is called the *sample size formula for estimating μ*. To use it, we plug in values for Z, σ, and E. For example, if Z = 1.28, σ = 5, and E = .4, what is the value of n?

$$n = \left(\frac{1.28(5)}{.4}\right)^2 = (16)^2 = 256$$

The solution value is n = 256, when Z = 1.28, σ = 5, and E = .4.

WORD PROBLEM EXAMPLE

A couple has saved $5,000, which is sitting in their bank savings account earning a 3.5% return. They would like to invest some of this money in a Treasury note that yields a 6% return. How much should they invest in the Treasury note and how much should they leave in their savings account so that the combined return in their two investments is 5%? (Assume no more money will be deposited into the savings account.)

Problem Solving Diagnostics:

1.	**Key words:**	Savings account, return, Treasury note, combined return
2.	**Numbers:**	$5,000 (total amount available for investment) 3.5% (return on savings account) 6% (return on Treasury note) 5% (combined return)
3.	**Picture:**	Not applicable
4.	**Question mark:**	Separate $5,000 into two amounts so that combined return from those two amounts is 5% or $250 (.05 times $5,000)

5. **Computations:** Let X be the amount in the savings account and let ($5,000 − X) be the amount invested in the Treasury note

Return on savings account = 3.5% of X
Return on Treasury note = 6% of ($5,000 − X)
Combined return = .035X + .06(5,000 − X)

The combined return must yield 5% of $5,000, or $250, so

$$.035X + .06(5,000 − X) = 250$$

Solve for X:

$$.035X + 300 − .06X = 250$$
$$.035X − .06X = 250 − 300$$
$$−.025X = −50$$
$$X = 2,000$$

The couple should leave $2,000 in their savings account (earning 3.5%) and invest $3,000 in a Treasury note (earning 6%). The overall return will be 5% or $250.

8.4 Solving an Inequality

In section 1.4 we reviewed inequalities (< and >) and provided guidelines for judging the larger or smaller of two numbers. Inequalities also can involve *expressions* like X, X + 3, or 2X + 7 in which the symbol X is a variable. Usually our interest in such inequalities is in finding which set of numbers for X (or whatever the variable symbol) make the inequality a true statement. For example, the inequality X + 3 > 7 is a true statement when X is 5, but it is a false statement when X is 2. In fact, the inequality X + 3 > 7 is true whenever X > 4.

Reducing the original inequality, X + 3 > 7, to the equivalent inequality X > 4 is called *solving the inequality*. As we learned earlier in solving equations, *solving an inequality for X* means we wish to isolate the variable X on one side of an inequality sign. There are several rules, given below, we must follow in solving inequalities. In these rules, the term <u>same sense</u> means the inequality symbol points in the same direction. The term <u>reverse sense</u> means the inequality symbol switches either from > to < or from < to >.

Solving an Equation

RULES
SOLVING INEQUALITIES

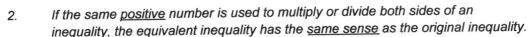

1. *If the same number is added to or subtracted from both sides of an inequality, the equivalent inequality has the <u>same sense</u> as the original inequality.*

2. *If the same <u>positive</u> number is used to multiply or divide both sides of an inequality, the equivalent inequality has the <u>same sense</u> as the original inequality.*

3. *If the same negative number is used to multiply or divide both sides of an inequality, the equivalent inequality has the <u>reverse sense</u> as the original inequality.*

Let us illustrate those rules with several examples. First, to solve the (original) inequality

$$X + 3 > 7$$

we subtract 3 from both sides. Rule 1 states that the resulting equivalent inequality will have the <u>same sense</u>:

$$X + 3 - 3 > 7 - 3$$
$$X > 4$$

For our second example, we will solve the (original) inequality:

$$5X - 9 < 1$$

As a first step we add 9 to both sides and retain the <u>same sense</u> of the inequality:

$$5X - 9 + 9 < 1 + 9$$
$$5X < 10$$

Then we divide both sides by 5. Rule 2 ensures the <u>same sense</u> of the resulting inequality:

$$\frac{5X}{5} < \frac{10}{5}$$

$$X < 2$$

In order for $5X - 9 < 1$ to be a true statement, we must have $X < 2$.

Finally, let us solve the inequality:

$$6 - 2X > 4$$

Subtract 6 from both sides of the inequality, retaining the <u>same sense</u> of the inequality:

$$6 - 2X - 6 > 4 - 6$$
$$-2X > -2$$

Now multiply both sides by $-\frac{1}{2}$. Rule 3 states that the equivalent inequality will have the <u>reverse sense</u>:

$$-2X > -2$$
$$(-\tfrac{1}{2}) \cdot (-2X) < (-\tfrac{1}{2}) \cdot (-2) \quad \text{(reverse sense)}$$
$$X < 1$$

This means that the original inequality $6 - 2X > 4$ is true only if $X < 1$.

SUMMARY EXAMPLE
Solving Inequalities

Solve the following inequalities for X:		(1) 7X – 3 > 4 (2) –½X > 5
Part (1)	Original inequality Add 3 to both sides (same sense) Divide both sides by 7 (same sense)	$7X - 3 > 4$ $7X - 3 + 3 > 4 + 3$ $7X > 7$ $\dfrac{7X}{7} > \dfrac{7}{7}$ $X > 1$ **Answer** ↗
Part (2)	Original inequality Multiply both sides by –2 (reverse sense)	$-\tfrac{1}{2}X > 5$ $-2(-\tfrac{1}{2}X) < -2(5)$ $X < -10$ **Answer** ↗

Solving an Equation

WORD PROBLEM EXAMPLE

A runner who normally runs in 10K road races has set her sights on running a marathon __and__ finishing the race in under 4 hours. A marathon is a race that covers 26.2 miles. What pace (approximately) must she run to reach her goal? ("Pace" means how fast, in minutes and seconds, it takes to run a mile.)

Problem Solving Diagnostics:

1.	**Key words:**	marathon, pace
2.	**Numbers:**	10K, or 10,000 kilometers (meaningless)
		4 hours (maximum time to complete race)
		26.2 miles (length of marathon)
3.	**Picture:**	Not applicable
4.	**Question mark:**	Find time, in minutes and seconds, to complete 26.2 miles in less than 4 hours
5.	**Computation:**	Let X = pace, in minutes
		4 hours = 240 minutes

If each mile is run in X minutes, then to complete the marathon in less than 4 hours, we must have:

$$\underbrace{X + X + X + X + \cdots + X + X}_{26 \text{ miles}} \; + \; \underset{+ .2 \text{ mile}}{.2X} < 240$$

$$26.2X < 240$$

Dividing both sides by the positive number 26.2 will retain the <u>same sense</u> of the inequality:

$$\frac{26.2X}{26.2} < \frac{240}{26.2}$$

$$X < 9.16$$

The decimal ".16" is about one-sixth of a minute ($\frac{1}{6} \approx .16$); one-sixth of a minute translates to about 10 seconds. Thus, to complete the marathon in under 4 hours, she should maintain a pace of 9 minutes and 10 seconds per mile.

Chapter 8 Exercises

8.1 Solve the following equation for X: $5X - 7 = 8$

8.2 Solve the following equation for Y:
$$2 = \frac{18}{-4Y + 1}$$

8.3 Solve the following equation for X: $2(5X - 2) - 3X = 4(3X - 3) - 7$

8.4 Solve the following equation for Y:
$$\frac{5}{Y + 2} = 2 - \frac{2Y - 8}{Y + 1}$$

8.5 Solve the following equation for X: $7X + X(5 - 3) = 8 + 3(2X - 1)$

8.6 Solve the following equation for μ:
$$Z = \frac{X - \mu}{\sigma}$$

8.7 Solve the following equation for σ:
$$Z = \frac{X - \mu}{\sigma}$$

IN EXERCISES 8.8 - 8.12 SOLVE THE EQUATION FOR THE SOLUTION VALUE OF THE UNKNOWN, GIVEN THE INDICATED NUMBERS

8.8 $X = -7, \mu = 1, \sigma = 2, Z = ?$

8.9 $X = 25, \mu = 10, Z = 2.5, \sigma = ?$

8.10 $\mu = 732, Z = -1.25, \sigma = 40, X = ?$

8.11 $X = 12.3, \sigma = .6, Z = 1.5, \mu = ?$

8.12 $Z = 1.85, X = -8, \mu = -8.629, \sigma = ?$

8.13 Solve the following equation for n:
$$E = Z\sqrt{\frac{\pi(1 - \pi)}{n}}$$

continued on the next page ...

Solving an Equation

Chapter 8 Exercises (continued)

8.14   Solve the following equation for Z. Then substitute π = .5 into the equation and express Z as a function of n and E.

$$E = Z\sqrt{\frac{\pi(1 - \pi)}{n}}$$

8.15   Solve the following equation for n:

$$E = Z \cdot \frac{\sigma}{\sqrt{n}} \cdot \sqrt{\frac{N - n}{N - 1}}$$

8.16   Solve the following inequality for X: $\frac{X - 10}{3} > 2$

8.17   Solve the following inequality for X: $2X + 4 > 3X - 8$

8.18   Solve the following inequality for X: $5 - 4X < -3$

8.19   Provided the service is good, Ryan always leaves a tip of 15% for his server in a restaurant. He bases the tip on the amount of his check <u>excluding</u> tax. If he left a $2 tip and if sales tax is 6%, how much was his total bill?

8.20   Samantha is saving to buy a new sound system. She gets paid weekly and from each pay check she saves $25. In three months (13 weeks) she will get a raise. After her raise she plans to double her savings to $50 per check. In how many weeks will she have enough to buy a $1,000 sound system? (Ignore the time value of money.)

8.21   Julie has a meeting at 2 p.m. in Charlotte. It is noon right now and she is in Greensboro, which is 90 miles from Charlotte. Interstate 85 connects the two cities. Assume that it will take her 10 minutes to park and walk to her meeting once she gets to Charlotte. If she plans to drive the speed limit -- 65 MPH -- on Interstate 85, what is the latest possible time she can leave for Charlotte and not arrive late for her appointment?

Solving an Equation

POSTTEST #8

Allow 20 Minutes
1 point each -- maximum score 7

QUESTIONS 8.1 - 8.3 REFER TO THE FOLLOWING EQUATION:
$$14 - .5X = 8$$

8.1 What is the coefficient of the variable X?

8.2 What is the solution value for the equation?

8.3 Does X = 4 satisfy the equation?

8.4 Solve the following equation for X:

$$\frac{15 - 14X}{2X + 5} + 7 = \frac{5}{4 - .8X}$$

8.5 Solve the following equation for n:

$$1.5 = \frac{2.5(6)}{\sqrt{n}}$$

8.6 Solve the following inequality for X:

$$-X + 6 < 2X - 3$$

8.7 Investments in certificates of deposit (CDS) produce a guaranteed return of 4%, while an investment in Treasury bonds will return 8%. How should you split an investment of $5,000 between CDS and Treasury bonds so as to achieve an overall return of 7%?

Solving an Equation

ANSWERS TO PRETESTS AND POSTTESTS

CHAPTER ONE

PRETEST ANSWERS

1.1	1.98
1.2	10.5%
1.3	.813
1.4	345.015
1.5	47.1%
1.6	$-1.23 > -1.32$
	(or $-1.32 < -1.23$)
1.7	7
1.8	$-3, -2, -1, 0, 1, 6$
	(low to high)
1.9	510,000
1.10	False.

POSTTEST ANSWERS

1.1	8
1.2	12.4%
1.3	.008
1.4	42.39
1.5	3.8%
1.6	$-9 > -9.1$
	(or $-9.1 < -9$)
1.7	62
1.8	Amount of U.S. dollars =
	$\dfrac{13}{.65} + \dfrac{77}{5.50} + \dfrac{20}{1.60}$
	$= 20.00 + 14.00 + 12.50$
	$= \$46.50$
1.9	$-7, -5, -4, -3, -1, 0$
	(low to high)
1.10	9,100

CHAPTER TWO

PRETEST ANSWERS

2.1	(i) no label for vertical axis
	(ii) numerical scaling for vertical axis is omitted
	(iii) no category labels for rectangles on horizontal axis
2.2	Bar graph
2.3	52.5%
2.4	135°
2.5	(c)

CHAPTER TWO

POSTTEST ANSWERS

2.1

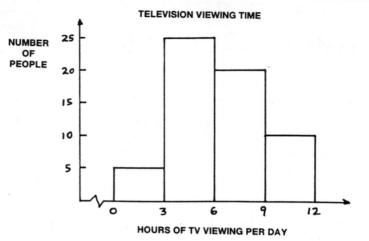

TELEVISION VIEWING TIME

2.2   Bivariate graph

2.3

Bar graph	Discrete Probability Distribution
(a) Whole numbers	Numbers less than 1
(b) Some gap space	No gaps

2.4   False

2.5   **(a)**   1.6
      (b)   4
      (c)   Yes

CHAPTER THREE

PRETEST ANSWERS

3.1   Greek letter sigma (Σ)
3.2   10
3.3   150
3.4   100
3.5   0
3.6   130
3.7 ΣX^2
3.8   39
3.9   2
3.10 -15

POSTTEST ANSWERS

3.1   **(a)** (10) Sum of the deviations
      (b) (9) Sum of the squared values
      (c) (1) Mean
      (d) (11) Sum of the squares for X
      (e) (6) Sum of the values squared
      (f) (8) Sum of the products
3.2   4
3.3   94
3.4   12
3.5 -36

CHAPTER FOUR

4.1   **(a)** x^7
      (b) x^{40}
      (c) x^{-5} or $\dfrac{1}{x^5}$

4.2   **(a)** $\sqrt{x}$
      (b) $\sqrt[3]{x}$
      (c) $\sqrt[5]{x^2}$

4.3   **(a)** 16,384
      (b) .0016
      (c) 1.4310

4.4   .00032

4.5   **(a)** 1.953×10^6
      (b) 5.77×10^{-4}

4.1   **(a)** x^{10}
      (b) x^{25}
      (c) x^{-10} or $\dfrac{1}{x^{10}}$

4.2   **(a)** $\sqrt[5]{8}$
      (b) $\sqrt[7]{x^3}$ or $(\sqrt[7]{x})^3$
      (c) $\dfrac{1}{\sqrt[3]{x}}$

4.3   **(a)** .0656
      (b) 730,200
      (c) 98,288,000

4.4   **(a)** 335.54432
      (b) 2.49805

4.5   1

CHAPTER FIVE

5.1   **(a)** 720
      (b) 6
      (c) 336

5.2   Permutations rule

5.3   Product rule:
      $10 \cdot 10 \cdot 10 = 1000$

5.4   Combinations rule:
      $_{35}C_5 = 324,632$

5.5   Permutations rule:
      $_3P_3 = 6$

5.6   Permutations rule:
      $_8P_2 = 56$

5.7   Combinations rule:
      $_{12}C_3 = 220$

5.8   Product rule:
      $12 \cdot 8 \cdot 3 = 288$

5.1   **(a)** 24
      (b) 1
      (c) $2.432902008 \times 10^{18}$
        (Note: shorter answers, such as
        2.433×10^{18} are also acceptable)
      (d) 21
      (e) 20
      (f) 5040

5.2   Permutations rule: $_5P_3 = 60$

5.3   Combinations rule: $_7C_3 = 35$

5.4   Product rule: $3 \cdot 3 \cdot 2 \cdot 2 = 36$

5.5   Permutations rule: $_6P_2 = 30$

CHAPTER SIX

PRETEST ANSWERS

6.1 The *sample space* is the collection of all possible outcomes (or sample points) for an experiment.

6.2 Use the product rule to find the total number of outcomes: $2 \cdot 2 \cdot 2 = 8$ outcomes
1st slot = penny   2nd slot = nickel   3rd slot = dime

Trio of letters represents the faces of each resting coin   (H = head, T = tail)

O_1 = HHH	O_3 = HTH	O_5 = HTT	O_7 = TTH
O_2 = HHT	O_4 = THH	O_6 = THT	O_8 = TTT

6.3 $E = \{O_5, O_6, O_7\}$

6.4 $P(E) = \frac{3}{8} = .375$

6.5 Use product rule to find the number of sample points in...

$$S: 2 \cdot 2 \cdot 2 \cdot 2 \cdot 2 = 32$$
$$E: 1 \cdot 1 \cdot 1 \cdot 1 \cdot 1 = 1$$
$$P(E) = \frac{1}{32} = .03125$$

POSTTEST ANSWERS

6.1 A *sample point* is the simplest result that is observable, but not 100% predictable, from an experiment.

6.2 Use the permutations rule to find the total number of outcomes: $_6P_2 = 30$ outcomes

Let   B = Buick   O = Oldsmobile   C = Cadillac
       P = Pontiac   V = Chevrolet   S = Saturn

O_1 = BC	O_6 = CB	O_{11} = VB	O_{16} = OB	O_{21} = PB	O_{26} = SB
O_2 = BV	O_7 = CV	O_{12} = VC	O_{17} = OC	O_{22} = PC	O_{27} = SC
O_3 = BO	O_8 = CO	O_{13} = VO	O_{18} = OV	O_{23} = PV	O_{28} = SV
O_4 = BP	O_9 = CP	O_{14} = VP	O_{19} = OP	O_{24} = PO	O_{29} = SO
O_5 = BS	O_{10} = CS	O_{15} = VS	O_{20} = OS	O_{25} = PS	O_{30} = SP

First letter is top ranked division.

6.3 Number of outcomes in event E is:

$_4P_2 = 12$: $E = \{O_2, O_3, O_4, O_{11}, O_{13}, O_{14}, O_{16}, O_{18}, O_{19}, O_{21}, O_{23}, O_{24}\}$

6.4 $P(E) = \frac{12}{30} = .4$

6.5 Use the product rule to find the number of sample points in ...

$$S: 1 \cdot 26 \cdot 26 = 676$$
$$E: 1 \cdot 1 \cdot 1 = 1$$

$$P(E) = \frac{1}{676} = .001479$$

NOTE: Because we know the first letter of the airport's code, there is only one choice for the first slot.

CHAPTER SEVEN

PRETEST ANSWERS

7.1
(a) $\{O_7\}$
(b) $\{O_1, O_4, O_5, O_6, O_7\}$
(c) $\{O_2, O_3\}$
(d) $\emptyset$

7.2
(a) 1/7
(b) 5/7
(c) 2/7
(d) 0

7.3
(a) $P(B^c) = 1 - P(B) = 1 - .60 = .40$
(b) $P(A \cup B) = P(A) +$
$P(B) - P(A \cap B)$
$= .30 + .60 - .20 = .70$
(c) $P(B \cap C) = P(C|B)\, P(B)$
$= .50(.60) = .30$

(d)

$$P(B|A) = \frac{P(A \cap B)}{P(A)} = \frac{.20}{.30} = .6667$$

(e) Yes, because $A \cap C = \emptyset$
(f) No, because $P(B)$ & $P(B|A)$ are not the same number. (Or because $P(A)$ and $P(A|B)$ are not the same number.)

POSTTEST ANSWERS

7.1 1

7.2
(a) $\{O_2, O_4, O_6, O_8\}$
(b) $\{O_1, O_3, O_5, O_6, O_7, O_9, O_{10}\}$
(c) $\{O_1, O_3, O_{10}\}$

7.3
(a) $P(E_3|E_2) = {}^2/_3 = .667$
(b) $P(E_1|E_3) = {}^3/_6 = .50$

7.4
(a) Yes, because $E_1 \cap E_2 = \emptyset$
(b) Yes, because $P(E_1)$ and $P(E_1|E_3)$ are the same number.

7.5 M = male, F = female, R = rock, C = country
(a) $P(M \cup R) = P(M) + P(R) - P(M \cap R)$
$= .60 + .35 - .20 = .75$

(b)

$$P(C|F) = \frac{P(C \cap F)}{P(F)} = \frac{.20}{.40} = .50$$

(c) $P(M \cap C) = P(C|M)$ times $P(M) = .50(.60) = .30$
(d) Yes, because $P(C)$ and $P(C|M)$ are the same number.

7.6
(a) .4

(b) $\dfrac{.3}{.5} = .6$

(c) .8

CHAPTER EIGHT

PRETEST ANSWERS

8.1 $X = 4$
8.2 $X = -4$
8.3 $X = 6$
8.4 $X = 100$
8.5 $X = 4Y + 6$
8.6 $X < 5$
8.7 3

POSTTEST ANSWERS

8.1 Coefficient: $-.5$
8.2 Solution value: $X = 12$
8.3 No
8.4 $X = 3.5$
8.5 $n = 100$
8.6 $X > 3$
8.7 $1,250 in CDs and $3,750 in Treasury bonds

Answers to Chapter 1 Exercises

1.1 1.0968142
1.2 31.6
1.3 .1945
1.4 22
1.5 66
1.6 186
1.7 23.2
1.8 **(a)** −.0270
    (b) −.0411
    (c) .1309
    (d) .0184
    (e) −.0079
1.9 **(a)** 5.7%
    (b) 18.0%
    (c) 70.11%
    (d) −21.6%
    (e) .5%
1.10 **(a)** 1.0962
    (b) 1.097
    (c) 1.10
1.11 **(a)** 3.30; 3.297
    (b) 11.03; 11.028
    (c) −5.10; −5.100
    (d) 2.24; 2.239
    (e) 2.27; 2.266
    (f) 0.58; 0.578
1.12 −32.43%
1.13 **(a)** 150%
    (b) −25%
1.14 **(a)** −4.67% **(c)** 0.00%
    (b) 11.20% **(d)** −150.00%
1.15 **(a)** 35
    (b) 1.73
    (c) 0.05
1.16 **(a)** >
    (b) >
    (c) <
    (d) <
    (e) <
    (f) <
1.17 **(a)** $52.12
    (b) $85.28

1.18 Key words: calories, grams of fat, hamburger, french fries, soft drink. Numbers and picture:

Item	Hamburger	French Fries	Soft Drink
Calories	342	400	75
% of fat Calories	50%	45%	0%
No. of fat calories	171	180	0

(To find number of fat calories, multiply calories by % of fat calories.)
Question mark: Find number of grams of fat.
Computations: Grams of fat = fat calories/9

	Hamburger	French Fries	Soft Drink
Fat calories	171	180	0
Grams of Fat	19	20	0

Grams of fat = 19 + 20 + 0 = 39.

1.19 8.25, 8.37, 8.38, 8.50 (low to high)
1.20 −6, −4, −2, −1, 0 (low to high)
1.21 **(a)** 75,546,530
    (b) 75,546,500
    (c) 75,547,000
    (d) 75,500,000
    (e) 76,000,000
1.22 **(a)** 901,660
    (b) 901,700
    (c) 902,000
    (d) 900,000
    (e) 1,000,000

Answers to Chapter 2 Exercises

2.1

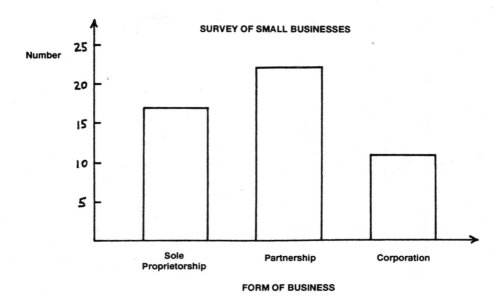

SURVEY OF SMALL BUSINESSES

2.2 The angles for each form of ownership are: partnership, 158°; sole proprietorship, 122°; and corporation, 80°.

SURVEY OF SMALL BUSINESSES

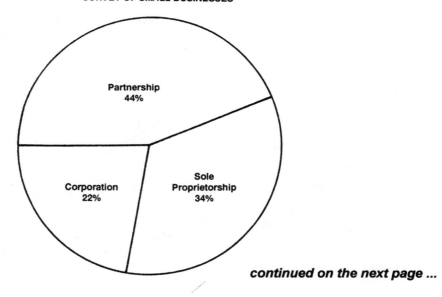

continued on the next page ...

2.3

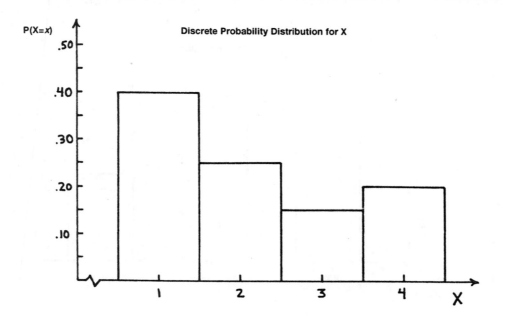

2.4 No, the line does not touch any of the points.

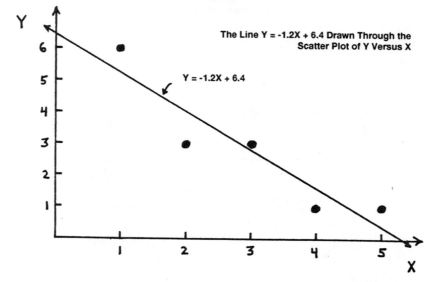

continued on the next page ...

Answers to Chapter 2 Exercises (cont.)

2.5

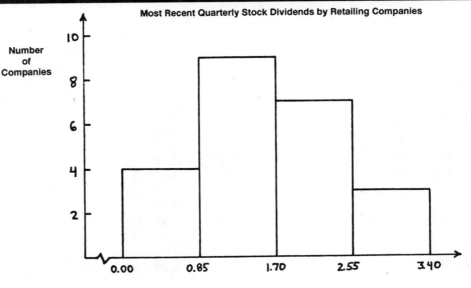

Most Recent Quarterly Stock Dividends by Retailing Companies

Number of Companies

Quarterly Dividends, in dollars and cents

2.6

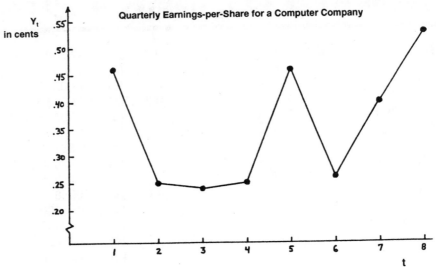

Quarterly Earnings-per-Share for a Computer Company

Y_t in cents

t

2.7 **(a)** bivariate graph
(b) bar graph
(c) bivariate graph
(d) bar graph

Answers to Chapter 3 Exercises

3.1   20
3.2   2
3.3   310
3.4   400
3.5   −50
3.6   270
3.7   **(a)** ΣX^2
      (b) $(\Sigma X)^2$
      (c) $\Sigma X - 10$
      (d) $\Sigma(X - 10)$
      (e) $\Sigma(X - 3)(Y - 2)$
3.8   −20
3.9   55
3.10   900
3.11   196
3.12   0
3.13   16

3.14   **(a)**   mean
      (b)   deviation
           (or deviation about the mean)
      (c)   sum of the products
      (d)   sum of squares for X
      (e)   sum of the squared values
3.15   1
3.16   .5
3.17   0
3.18   18
3.19   0
3.20   28
3.21   −3
3.22   **(a)** ΣM
      (b) ΣfM^2
      (c) $(\Sigma fM)^2$
3.23   **(a)** $50   **(b)**   132

Answers to Chapter 4 Exercises

4.1   **(a)** x^{11} **(d)** x^{24}
      (b) x^{10} **(e)** x^3
      (c) x^2 **(f)** x
4.2   **(a)** $\sqrt[3]{x}$
      (b) $\sqrt[9]{x^2}$
      (c) $\dfrac{1}{\sqrt{x}}$
      (d) $\sqrt[5]{x^6}$
4.3   **(a)** x^{-2}
      (b) x^{-3}
      (c) $x^{-\frac{1}{2}}$
4.4   **(a)** $\sqrt[3]{7}$
      (b) $\sqrt[5]{r^2}$
      (c) $\sqrt[n]{21}$
4.5   **(a)**   1.913
      (b)   1.741
      (c)   1.463

4.6   **(a)**   32,768   **(d)**   627.2220
      (b)   .0109   **(e)**   1.9120
      (c)   1.0051   **(f)**   8.5460
4.7   **(a)** 7.9268×10^4
      (b) 8.903274×10^2
      (c) 1.52399025×10^8
      (d) 6.561×10^{-1}
      (e) 9.426×10^{-9}
      (f) 1×10^0
4.8   **(a)**   .0088
      (b)   109,500
      (c)   .00000733
      (d)   2154.23
      (e)   334,310
      (f)   129,700,000,000
4.9 $\Sigma fM^2 = 780$; $\Sigma f(M - 8)^2 = 140$
4.10 $\Sigma x^2 P(X = x) = 7.4$
4.11   Calculator work (answers provided in the exercise).

Answers to Chapter 5 Exercises

5.1   **(a)**   24
        (b)   1
        (c)   40,320

5.2   **(a)** $2.092278989 \times 10^{13}$ *(NOTE: shorter answers, such as 2.092×10^{13} are also acceptable.)*
        (b) $2.631308369 \times 10^{35}$ (or 2.631×10^{35})
        (c) $1.268869322 \times 10^{89}$ (or 1.269×10^{89})

5.3   **(a)**   45
        (b)   286
        (c)   15,120

5.4   **(a)** $_8C_4 = 70$
        (b) $_{15}C_3 = 455$
        (c) $_{52}C_5 = 2{,}598{,}960$
        (d) $_4C_2 = 6$
        (e) $_{15}C_5 = 3{,}003$

5.5.   **(a)** $_8P_4 = 1{,}680$
        (b) $_{15}P_3 = 2{,}730$
        (c) $_{52}P_5 = 311{,}875{,}200$
        (d) $_4P_2 = 12$
        (e) $_{15}P_5 = 360{,}360$

5.6   **(a)** $26 \cdot 26 \cdot 26 \cdot 10 \cdot 10 \cdot 10 = 17{,}576{,}000$
        (b) $6 \cdot 8 \cdot 10 = 480$
        (c) $5 \cdot 3 \cdot 4 \cdot 2 = 120$
        (d) $3 \cdot 3 \cdot 3 = 27$
        (e) $7 \cdot 8 \cdot 3 \cdot 4 \cdot 5 = 3{,}360$

5.7   **(a)** $26 \cdot 21 \cdot 26 \cdot 10 \cdot 10 \cdot 10 = 14{,}196{,}000$
        (b) $3 \cdot 8 \cdot 10 = 240$ (first slot contains female chair from the smallest department)
        (c) $5 \cdot 9 = 45$ (first slot corresponds to airlines sector; second slot combines the other 3 sectors)
        (d) $3 \cdot 4 \cdot 4 = 48$ (first digit cannot be the digit 0)
        (e)   Let B = bluegrass, C = country, J = jazz, R = rock 'n roll, and N = new age. Number of pairs is $_5C_2 = 10$. Following are the possibilities for each pair. Total is found by adding all possibilities:

BC: $7 \cdot 8 = 56$	BN: $7 \cdot 5 = 35$	CN: $8 \cdot 5 = 40$
BJ: $7 \cdot 3 = 21$	CJ: $8 \cdot 3 = 24$	JR: $3 \cdot 4 = 12$
BR: $7 \cdot 4 = 28$	CR: $8 \cdot 4 = 32$	JN: $3 \cdot 5 = 15$
	RN: $4 \cdot 5 = 20$	

                                                                 TOTAL = 283

5.8   **(a)**   Combinations rule: $_{40}C_6 = 3{,}838{,}380$
        (b)   Combinations rule: $_{42}C_6 = 5{,}245{,}786$
        (c)   Combinations rule: $_{44}C_6 = 7{,}059{,}052$

continued on the next page ...

Answers to Chapter 5 Exercises (cont.)

5.9   Combinations rule: $_{52}C_6 = 20,358,520$

5.10   Permutations rule: total numbers of ways is $_5P_5 = 120$. Only 1 correct way; 119 incorrect ways.

5.11   Combinations rule: $_7C_4 = 35$

5.12   **(a)**   Combinations rule: $_3C_2 = 3$

          (b)   Combinations rule: $_4C_2 = 6$

          (c)   Product rule: $3 \cdot 6 = 18$ (think of the first "slot" as the pair of women and the second "slot" as the pair of men).

5.13   Permutations rule: $_{10}P_3 = 720$

5.14   Combinations rule: $_{10}C_3 = 120$ (or $_{10}C_7$, which is also 120)

5.15   Product rule: $8 \cdot 2 \cdot 9 = 144$

Answers to Chapter 6 Exercises

6.1   Use product rule to find total number of outcomes:

$$\underline{6} \cdot \underline{6} = 36 \text{ outcomes}$$

1st slot = white die   2nd slot = red die

$O_1 = (1,1)$	$O_{13} = (3,1)$	$O_{25} = (5,1)$
$O_2 = (1,2)$	$O_{14} = (3,2)$	$O_{26} = (5,2)$
$O_3 = (1,3)$	$O_{15} = (3,3)$	$O_{27} = (5,3)$
$O_4 = (1,4)$	$O_{16} = (3,4)$	$O_{28} = (5,4)$
$O_5 = (1,5)$	$O_{17} = (3,5)$	$O_{29} = (5,5)$
$O_6 = (1,6)$	$O_{18} = (3,6)$	$O_{30} = (5,6)$
$O_7 = (2,1)$	$O_{19} = (4,1)$	$O_{31} = (6,1)$
$O_8 = (2,2)$	$O_{20} = (4,2)$	$O_{32} = (6,2)$
$O_9 = (2,3)$	$O_{21} = (4,3)$	$O_{33} = (6,3)$
$O_{10} = (2,4)$	$O_{22} = (4,4)$	$O_{34} = (6,4)$
$O_{11} = (2,5)$	$O_{23} = (4,5)$	$O_{35} = (6,5)$
$O_{12} = (2,6)$	$O_{24} = (4,6)$	$O_{36} = (6,6)$

6.2   Use product rule to find total number of outcomes:

$$\underline{3} \cdot \underline{3} = 9 \text{ outcomes}$$

1st slot = first day   2nd slot = second day

$O_1 = (U,U)$	$O_4 = (D,U)$	$O_7 = (S,U)$
$O_2 = (U,D)$	$O_5 = (D,D)$	$O_8 = (S,D)$
$O_3 = (U,S)$	$O_6 = (D,S)$	$O_9 = (S,S)$

continued on the next page ...

6.3 Use combinations rule to find total number of outcomes:

$$_5C_3 = 10 \text{ outcomes}$$

Trio of letters represents elected candidates
(A = Adams, B = Becker, C = Chou,
D = Davis, E = Espinosa)

O_1 = (ABC)	O_5 = (ACE)	O_9 = (BDE)
O_2 = (ABD)	O_6 = (ADE)	O_{10} = (CDE)
O_3 = (ABE)	O_7 = (BCD)	
O_4 = (ACD)	O_8 = (BCE)	

6.4 Use product rule to find total number of outcomes:

$$\underline{3} \cdot \underline{3} = 9 \text{ outcomes}$$

1st slot = first marble 2nd slot = second marble

Pair of letters represents colors of selected marbles (R = red,
B = blue, O = orange)

. O_1	. O_4	. O_7	O_1 = RR	O_6 = BO
. O_2	. O_5	. O_8	O_2 = RB	O_7 = OR
. O_3	. O_6	. O_9 S	O_3 = RO	O_8 = OB
			O_4 = BR	O_9 = OO
			O_5 = BB	

6.5 **(a)** $E_1 = \{O_4, O_{10}, O_{16}, O_{22}, O_{28}, O_{34}\}$; $P(E_1) = 6/36 = .1667$

(b) $E_2 = \{O_7, O_9, O_{11}\}$; $P(E_2) = 3/36 = .0833$

(c) $E_3 = \{O_{12}, O_{17}, O_{22}, O_{27}, O_{32}\}$; $P(E_3) = 5/36 = .1389$

6.6 **(a)** $E_1 = \{O_2, O_3, O_6, O_9\}$; $P(E_1) = 4/10 = .4$

(b) $E_2 = \{O_1, O_4, O_5, O_7, O_8, O_{10}\}$; $P(E_2) = 6/10 = .6$

(c) $E_3 = \{O_4, O_5, O_6, O_{10}\}$; $P(E_3) = 4/10 = .4$

(d) $E_4 = \{O_3, O_5, O_6\}$; $P(E_4) = 3/10 = .3$

continued on the next page...

6.7 **(a)** $E_1 = \{O_9\}$; $P(E_1) = 1/9 = .1111$

(b) $E_2 = \{O_1, O_3, O_7, O_9\}$; $P(E_2) = 4/9 = .4444$

(c) $E_3 = \{O_2, O_3, O_4, O_6, O_7, O_8\}$; $P(E_3) = 6/9 = .6667$

6.8 Use combinations rule to find number of sample points in...

S: $_{52}C_5 = 2,598,960$

E: $_{26}C_5 = 65,780$

$$P(E) = \frac{65780}{2598960} = .0253$$

6.9 Use product rule to find number of sample points in...

S: $\underline{26} \cdot \underline{26} \cdot \underline{26} \cdot \underline{10} \cdot \underline{10} \cdot \underline{10} = 17,576,000$

E: $\underline{1} \cdot \underline{26} \cdot \underline{26} \cdot \underline{10} \cdot \underline{10} \cdot \underline{1} = 67,600$

(first slot must have letter A; last slot must have digit 9)

$$P(E) = \frac{67600}{17576000} = .003846$$

6.10 Use combinations rule to find number of sample points in...

S: $_{10}C_3 = 120$

E: $_5C_3 = 10$

$$P(E) = \frac{10}{120} = .0833$$

Answers to Chapter 7 Exercises

7.1   **(a)** $\{O_1, O_2\}$
        (b) $\{O_1, O_2, O_5, O_6, O_8\}$
        (c) $\{O_5, O_6, O_8\}$
        (d) $\{O_2\}$
        (e) $\{O_8\}$
        (f) $\{O_1, O_3, O_4, O_5, O_6, O_7, O_8\}$
        (g)   No, because E_1 and E_2 share a sample point (O_8) in common.

7.2   **(a)** $\{O_2, O_4, O_8, O_9, O_{10}, O_{11}\}$
        (b) $\{O_1, O_7\}$
        (c) $\{O_4, O_8\}$
        (d) $\{O_2, O_4, O_8, O_9, O_{10}, O_{11}, O_{12}\}$
        (e) $\{O_1, O_3, O_5, O_6, O_7, O_9, O_{10}\}$
        (f) $\{O_3, O_5, O_6, O_9, O_{10}\}$
        (g)   Yes, because $E_1 \cap E_2 = \varnothing$
        (h)   Yes, because $E_3 \cap E_4 = \varnothing$

7.3   **(a)** $P(E_3^c) = 2/8 = .25$
        (b) $P(E_1 \cup E_2) = 5/8 = .625$
        (c) $P(E_2 \cap E_3) = 3/8 = .375$
        (d) $P(E_3^c \cap E_2) = 1/8 = .125$
        (e) $P(E_1 \cap E_2 \cap E_3) = 1/8 = .125$
        (f) $P(E_2^c \cup E_3) = 7/8 = .875$

   7.4   **(a)** $P(E_1 \cup E_4) = 6/12 = .50$
        (b) $P(E_2 \cap E_3) = 2/12 = .1667$
        (c) $P(E_1 \cap E_3) = 2/12 = .1667$
        (d) $P(E_2^c) = 7/12 = .5833$
        (e) $P(E_2 \cup E_4) = 7/12 = .5833$
        (f) $P(E_1^c \cap E_3^c) = 5/12 = .4167$

7.5   **(a)** $P(E_1|E_3) = 1/6 = .1667$
        (b) $P(E_3|E_2) = 3/4 = .75$
        (c)   No, because $P(E_1)$ and $P(E_1|E_3)$ are not the same number.

7.6   **(a)** $P(E_2|E_4) = 0$
        (b) $P(E_3|E_1) = 2/4 = .50$
        (c) $P(E_2|E_3) = 2/5 = .40$
        (d)   No, because $P(E_3)$ and $P(E_3|E_1)$ are not the same number.
        (e)   No, because $P(E_2)$ and $P(E_2|E_4)$ are not the same number.

continued on the next page...

Answers to Chapter 7 Exercises (cont.)

7.7 (a) $A \cup B$

(b) $(A \cup B)^c$ (or $A^c \cap B^c$)

(c) $A \cap C \cap B^c$

(d) $A \cap B$

7.8 (a) $P(A \cup B) = P(A) + P(B) = .10 + .20 = .30$

(b) $P(A \cup C) = P(A) + P(C) - P(A \cap C) = .10 + .60 - .03 = .67$

(c) $P(B \cap C) = P(B|C) P(C) = (.25)(.60) = .15$

(d) $P(B|A) = \dfrac{P(A \cap B)}{P(A)} = \dfrac{0}{.10} = 0$

(e) $P(A|C) = \dfrac{P(A \cap C)}{P(C)} = \dfrac{.03}{.60} = .05$

(f) $P(C|A) = \dfrac{P(A \cap C)}{P(A)} = \dfrac{.03}{.10} = .30$

(g) No, because P(B) and P(B|A) are not the same number.

(h) No, because P(A) and P(A|C) are not the same number (or because P(C) and P(C|A) are not the same number).

7.9 (a) $P(A \cup D) = P(A) + P(D) - P(A \cap D) = .80 + .35 - .30 = .85$

(b) $P(B|C) = \dfrac{P(B \cap C)}{P(C)} = \dfrac{.05}{.25} = .20$

(c) $P(A \cap C) = P(C|A) P(A) = (.25)(.80) = .20$

(d) $P(D|A) = \dfrac{P(A \cap D)}{P(A)} = \dfrac{.30}{.80} = .375$

(e) From $P(B \cup C) = P(B) + P(C) - P(B \cap C)$, we substitute the known probabilities and solve for P(B): $.40 = P(B) + .25 - .05 \Rightarrow P(B) = .20$

(f) $P(B \cap D) = P(D|B) P(B) = (.25)(.20) = .05$

(g) Yes, because P(C) and P(C|A) are the same number.

(h) No, because P(D) and P(D|A) are not the same number.

(i) Yes, because P(B) and P(B|C) are the same number.

continued on the next page...

7.10 **(a)** $P(M) = 1 - P(F) = 1 - .50 = .50$

(b) $P(L) = .10$

(c) $P(F \cap L) = 0$

(d) $P(F|A) = \dfrac{P(F \cap A)}{P(A)} = \dfrac{.05}{.05} = 1$

(e) $P(L|M) = \dfrac{P(L \cap M)}{P(M)} = \dfrac{.10}{.50} = .20$

(f) $P(R|F) = \dfrac{P(R \cap F)}{P(F)} = \dfrac{.45}{.50} = .90$

7.11 **(a)** .40

(b) .10

(c) .20

7.12 **(a)** $P(B) = .20$

(b) $P(S) = .77$

(c) $P(B \cap N) = .18$

(d) $P(A|S) = \dfrac{P(A \cap S)}{P(S)} = \dfrac{.75}{.77} = .974$

(e) $P(N|B) = \dfrac{P(N \cap B)}{P(B)} = \dfrac{.18}{.20} = .90$

(f) $P(S|A)$

7.13 V = vegetarian, M = meat eater, Y = young, MA = middle-aged, S = senior

(a) $P(M) = .90$

(b) $P(Y^c) = 1 - P(Y) = 1 - .40 = .60$

(c) $P(MA|V) = P(MA)$, because the events "middle-aged" and "vegetarian" are independent events. Thus, $P(MA|V) = .40$

(d) $P(Y \cap V) = P(Y|V)P(V) = (.50)(.10) = .05$

(e) $P(S \cup M) = P(S) + P(M) - P(S \cap M)$
$$= .20 + .90 - .19$$
$$= .91$$

Answers to Chapter 8 Exercises

8.1 $X = 3$

8.2 $Y = -2$

8.3 $X = 3$

8.4 $Y = -3$

8.5 $X = \dfrac{5}{3} = 1.666...$

8.6 $\mu = X - Z\sigma$

8.7 $\sigma = \dfrac{X - \mu}{Z}$

8.8 $Z = -4$

8.9 $\sigma = 6$

8.10 $X = 682$

8.11 $\mu = 11.4$

8.12 $\sigma = .34$

8.13 $n = \dfrac{Z^2 \pi(1 - \pi)}{E^2}$

8.14

$$Z = \dfrac{E}{\sqrt{\dfrac{\pi(1 - \pi)}{n}}}$$

$$Z = \dfrac{E\sqrt{n}}{.5} = 2E\sqrt{n}$$

8.15 $n = \dfrac{NZ^2 \sigma^2}{E^2(N - 1) + Z^2 \sigma^2}$

8.16 $X > 16$

8.17 $X < 12$

8.18 $X > 2$

8.19 **Key words:**

Tip, bill, sales tax

Numbers:

15% (tip percent on meal excluding tax)
6% (sales tax); $2 (tip amount)

Picture: Not applicable

Question mark:

Find total amount of bill

Computations:

Let X = total bill, then X = amount for food & drink + tax

→ Amount for food & drink can be found from tip amount:

→ tip amount = 15% of amount for food & drink
$ 2.00 = .15 (amount)
$13.33 = amount

→ Tax is found by taking 6% of amount for food & drink, so the total bill is

$X = $13.33 + 6\%$ of $13.33
$= $13.33 + .80 = \underline{\mathbf{\$14.13}}$
Answer ↗

continued on the next page...

Answers to Chapter 8 Exercises (cont.)

8.20　**Key words:**　Weekly, raise, double her savings

　　　　　Numbers:　$25 (weekly savings <u>before</u> raise)

　　　　　　　　　　　$50 (weekly savings <u>after</u> raise)

　　　　　　　　　　　$1,000 (cost of sound system)

　　　　　Picture:　Not applicable

　　　　　Question mark:

Determine number of weeks until savings exceeds (or equals) $1,000.

Computations:

She will not have enough money in savings <u>before</u> her raise because 13 weeks times $25 is only $325, which is well short of the $1,000 needed.

→ Let X = the number of weeks of saving $50 per check.

→ Solve the following inequality for X:

　　$325 + 50X > $1,000

　　50X > 675

　　X > 13.5

The figure of 13.5 means thirteen-and-a-half weeks of saving $50 per week. But because she gets paid weekly, she won't have enough until 14 weeks have elapsed <u>after</u> her raise. In total, it will take her 13 weeks before her raise and 14 weeks after, or

<u>27 weeks</u>. ←Answer

8.21　**Key words:**　Greensboro ... 90 miles from Charlotte, park and walk, speed limit, not arrive late

　　　　　Numbers:　90 miles (distance between cities)

　　　　　　　　　　　10 minutes (time to park and walk)

　　　　　　　　　　　65 MPH (speed limit)

　　　　　　　　　　　12:00 noon (current time)

　　　　　　　　　　　2:00 p.m. (meeting time)

　　　　　　　　　　　85 (Interstate number - meaningless)

　　　　　Picture:　•..•++++++++++++++++ X

Greensboro　　← 90 miles →　　Charlotte　← 10 minutes → Meeting site

　　　　　Question:　Find the latest departure time from Greensboro to ensure she does not arrive late at the meeting site.

　　　　　Computation:　Time to traverse the distance between Greensboro and Charlotte is the key. How many minutes will it take to travel 90 miles if she travels at a rate of 65 MPH? In hours, the answer is:

$$\frac{90}{65} = 1.385 \text{ hours}$$

The decimal .385 hours translates to approximately 23 minutes:

.385 times 60 ≈ 23 minutes

The total travel time to the meeting site is:

1 hour and 23 minutes + 10 minutes, or 1 hour and 33 minutes. The latest possible departure time is **12:27 p.m.** (Assuming no traffic tie-ups!)　　　　　**Answer** ↗

DEFINITIONS

NAME	PAGE

«FORMULAS»

RULES

INDEX (continued)

TERMS

Terms continued on the next page.....

TERMS (continued)

SYMBOLS

(listed in the order in which they appear in the book)